COMSOL® FOR ENGINEERS

LICENSE, DISCLAIMER OF LIABILITY, AND LIMITED WARRANTY

By purchasing or using this book (the "Work"), you agree that this license grants permission to use the contents contained herein, but does not give you the right of ownership to any of the textual content in the book or ownership to any of the information or products contained in it. *This license does not permit uploading of the Work onto the Internet or on a network (of any kind) without the written consent of the Publisher.* Duplication or dissemination of any text, code, simulations, images, etc. contained herein is limited to and subject to licensing terms for the respective products, and permission must be obtained from the Publisher or the owner of the content, etc., in order to reproduce or network any portion of the textual material (in any media) that is contained in the Work.

MERCURY LEARNING AND INFORMATION ("MLI" or "the Publisher") and anyone involved in the creation, writing, or production of the companion disc, accompanying algorithms, code, or computer programs ("the software"), and any accompanying Web site or software of the Work, cannot and do not warrant the performance or results that might be obtained by using the contents of the Work. The author, developers, and the Publisher have used their best efforts to insure the accuracy and functionality of the textual material and/or programs contained in this package; we, however, make no warranty of any kind, express or implied, regarding the performance of these contents or programs. The Work is sold "as is" without warranty (except for defective materials used in manufacturing the book or due to faulty workmanship).

The author, developers, and the publisher of any accompanying content, and anyone involved in the composition, production, and manufacturing of this work will not be liable for damages of any kind arising out of the use of (or the inability to use) the algorithms, source code, computer programs, or textual material contained in this publication. This includes, but is not limited to, loss of revenue or profit, or other incidental, physical, or consequential damages arising out of the use of this Work.

The sole remedy in the event of a claim of any kind is expressly limited to replacement of the book, and only at the discretion of the Publisher. The use of "implied warranty" and certain "exclusions" vary from state to state, and might not apply to the purchaser of this product.

COMSOL® FOR ENGINEERS

Mehrzad Tabatabaian

MERCURY LEARNING AND INFORMATION
Dulles, Virginia
Boston, Massachusetts
New Delhi

Copyright ©2014 by MERCURY LEARNING AND INFORMATION. All rights reserved.

This publication, portions of it, or any accompanying software may not be reproduced in any way, stored in a retrieval system of any type, or transmitted by any means, media, electronic display or mechanical display, including, but not limited to, photocopy, recording, Internet postings, or scanning, without prior permission in writing from the publisher.

Publisher: David Pallai
MERCURY LEARNING AND INFORMATION
22841 Quicksilver Drive
Dulles, VA 20166
info@merclearning.com
www.merclearning.com
1-800-758-3756

This book is printed on acid-free paper.

M. Tabatabaian. *COMSOL® for Engineers.*

ISBN: 978-1-938549-53-3

The publisher recognizes and respects all marks used by companies, manufacturers, and developers as a means to distinguish their products. All brand names and product names mentioned in this book are trademarks or service marks of their respective companies. Any omission or misuse (of any kind) of service marks or trademarks, etc. is not an attempt to infringe on the property of others.

Library of Congress Control Number: 2013958000

141516321

Our titles are available for adoption, license, or bulk purchase by institutions, corporations, etc. For additional information, please contact the Customer Service Dept. at 1-800-758-3756 (toll free).

The sole obligation of MERCURY LEARNING AND INFORMATION to the purchaser is to replace the disc, based on defective materials or faulty workmanship, but not based on the operation or functionality of the product.

*To the memories of my father,
whose support and guidance are still felt to this day.
To my family, for their support and encouragement.*

CONTENTS

Preface — xi

Chapter 1: Introduction — 1

Chapter 2: Finite Element Method—A Summary — 5

    Overview — 5

    FEM Formulation — 8

    Matrix Approach — 9

        Example 2.1: Analysis of a 2D Truss — 9

        General Procedure for Global Matrix Assembly — 13

        Example 2.2: Global Matrix for Triangular Elements — 14

    Weighted Residual Approach — 15

    Galerkin Method — 15

    Shape Functions — 16

    Convergence and Stability — 17

        Example 2.3: Heat Transfer in a Slender Steel Bar — 18

    Exercise Problems — 21

    References — 21

Chapter 3: COMSOL—A Modeling Tool for Engineers — 23

- Overview — 23
- COMSOL Interface — 24
- COMSOL Modules — 32
- COMSOL Model Library and Tutorials — 33
- General Guidelines for Building a Model — 34

Chapter 4: COMSOL Models for Physical Systems — 37

- Overview — 37
- **Section 4.1: Static and Dynamic Analysis of Structures** — 38
  - Example 4.1: Stress Analysis for a Thin Plate Under Stationary Loads — 38
  - Example 4.2: Dynamic Analysis for a Thin Plate: Eigenvalues and Modal Shapes — 49
  - Example 4.3: Parametric Study for a Bracket Assembly: 3D Stress Analysis — 53
  - Example 4.4: Buckling of a Column with Triangular Cross-section: Linearized Buckling Analysis — 67
  - Example 4.5: Static and Dynamic Analysis for a 2D Bridge-support Truss — 76
  - Example 4.6: Static and Dynamic Analysis for a 3D Truss Tower — 88
- **Section 4.2: Dynamic Analysis and Models of Internal Flows: Laminar and Turbulent** — 95
  - Example 4.7: Axisymmetric Flow in a Nozzle: Simplified Water-jet — 96
  - Example 4.8: Swirl Flow Around a Rotating Disk: Laminar Flow — 105
  - Example 4.9: Swirl Flow Around a Rotating Disk: Turbulent Flow — 114
  - Example 4.10: Flow in a U-shape Pipe with Square Cross-sectional Area: Laminar Flow — 118
  - Example 4.11: Double-driven Cavity Flow: Moving Boundary Conditions — 129

Example 4.12: Water Hammer Model: Transient Flow Analysis	142
Example 4.13: Static Fluid Mixer Model	150
Section 4.3: Modeling of Steady and Unsteady Heat Transfer in Media	**159**
Example 4.14: Heat Transfer in a Multilayer Sphere	159
Example 4.15: Heat Transfer in a Hexagonal Fin	165
Example 4.16: Transient Heat Transfer Through a Nonprismatic Fin with Convective Cooling	173
Example 4.17: Heat Conduction Through a Multilayer Wall with Contact Resistance	180
Section 4.4: Modeling of Electrical Circuits	**185**
Example 4.18: Modeling an RC Electrical Circuit	185
Example 4.19: Modeling an RLC Electrical Circuit	188
Section 4.5: Modeling Complex and Multiphysics Problems	**193**
Example 4.20: Stress Analysis for an Orthotropic Thin Plate	194
Example 4.21: Thermal Stress Analysis and Transient Response of a Bracket	197
Example 4.22: Static Fluid Mixer with Flexible Baffles	205
Example 4.23: Double Pendulum: Multibody Dynamics	214
Example 4.24: Multiphysics Model for Thermoelectric Modules	219
Example 4.25: Acoustic Pressure Wave Propagation in an Automotive Muffler	228
Exercise Problems	**238**
References	**243**
Suggested Further Readings	**244**
Trademark References	**244**
Index	**245**

PREFACE

This book is written for engineers, engineering students, and other practitioners in engineering fields. The main objective of the material is to introduce and help readers to use COMSOL® as an engineering tool for modeling by solving examples that either directly could be used or could become a guide for modeling similar or more complicated problems. It would be exhaustive to include all features available in COMSOL in a single book; our objective is to provide a collection of examples and modeling guidelines through which readers could build their own models.

Readers are assumed to know or at least be familiar with the principles of numerical modeling and finite element method (FEM). We took a *flexible-level* approach for presenting the materials along with using practical examples. The mathematical fundamentals, engineering principles, and design criteria are presented as integral parts of examples. At the end of each chapter we have added references that contain more in-depth physics, technical information, and data; these are referred to throughout the book and used in the examples. This approach allows readers to learn the materials at their desired level of complexity.

COMSOL for Engineers could be used as a textbook complementing another text that provides background training in engineering computations and methods, such as FEM. Examples provided in this book should be considered as "lessons" for which background physics could be explained in more detail. Exercise problems, or their variations, could be used for homework assignments.

We start each chapter with an overview, background physics, and mathematical models to set the foundation. We then present the relevant modeling techniques and materials through several examples. The examples progress from simple to more complex and are designed to complement one another, where applicable. Several exercise questions are provided following and relevant to each example. We use the COMSOL software tool (version 4.3 series[1]) for solving the examples. Where suitable, we also compare the modeling results with existing analytical, experimental, or other relevant models. Detailed steps are provided (relevant to version 4.3) to build the relevant model for each example, but it is recommended that readers, especially students, go through all models to master applications of COMSOL. The purpose of using COMSOL software is to introduce this tool to engineering students, engineers, and researchers.

This book is composed of the following chapters:

Chapter 1: Introduction

In this chapter, we discuss why multiphysics modeling is becoming a necessary tool for engineering design and analysis in modern engineering education and practices.

Chapter 2: Finite Element Method—A summary

In this chapter, we provide a summary of FEM and its main merits. This is intended to help the reader understand some technical features of COMSOL. It also provides a common level of understanding of this popular and powerful engineering computational method for readers with different educational backgrounds.

Chapter 3: COMSOL—A Modeling Tool for Engineers

In this chapter, we introduce the main features and structure of COMSOL (version 4.3), including modules available and main references for further readings for interested readers. Additional details for using this modeling tool are provided when it is used for solving examples later in the book.

Chapter 4: COMSOL Models for Physical Systems

In this chapter, we use COMSOL to solve examples that represent "practical" engineering problems involving fluids, solids, and electrical networks. Several examples and step-by-step instructions to build the models in COMSOL and interpretation of results are presented. Readers will find it useful to understand the preceding chapters before attempting the content in Chapter 4.

---

[1] http://www.comsol.com/support/releasehistory/

During publication of this book, a new version of COMSOL (version 4.4) was announced. The models available from the accompanying CD could be used with the new software version.

M. Tabatabaian
Vancouver, BC
February, 2014

CHAPTER 1

INTRODUCTION

Engineering practice foundations are mathematical models, physics principles, and empirical results obtained from experiments for defining design criteria. When we mention physics in this book we mean overall science, which includes all disciplines such as chemistry, biology, etc. An engineer should know the Laws of Physics very well and use the relevant mathematical models and their solutions, either exact or numerical, in practice to design parts, systems, and complex machines that work and function with certain reliability for an assumed lifetime.

Knowing the exact behavior of a complete system or a system component under actual "real life" type loads is an extremely difficult task. Hence engineers use a safety/design factor to overcome the probability of weaknesses and defects in materials used or extreme loads applied to a system or component under extreme conditions. For example, an engineer who designs a curved beam for a given load assumes that the material is free from defects and the load does not exceed the corresponding design limits. However, that beam might be subject to extreme dynamic loads due to vibration resulting from a strong earthquake, and subsequently fail. Designing for extreme cases is "possible" but is neither practical nor economical. Safety factors should be considered in accordance with design codes, which usually address the minimum requirements. To assist in this task, an engineer may use modeling tools to simulate the behavior of the curved beam such that the beam's strength is sufficient to resist the real-world loads applied to it. The modeling task and application of software tools are becoming more common in modern engineering practices, as shown in Figure 1.1. Modeling results can support optimization and refinement of a design before the physical prototype is built and minimize the duration of the design process. In addition, application of modeling helps to minimize the final cost of a prototype or product.

A definition of a "model" seems useful at this point. Actually, modeling has a long history that begins in ancient times when scientists used "equations" to relate variables or parameters to one another (e.g., Archimedes, Tales, Khwarizmi). Later, scientists and mathematicians developed "equations" that could represent the way that natural phenomena work and materials behave. These "equations" are sometimes referred to as *Laws of Physics* and *constitutive equations*, because they have been validated over time and the obtained results match what we experience or measure in the real world (of course, with some approximations). For example, Newton's second law is given as a "model" that predicts the behavior of materials under given forces applied to them. In other words, it is a relationship between forces applied to a material point (or a body mass) and the change of its momentum with respect to time. Similarly, Ohm's law is a model that relates the voltage applied across a resistor to the electrical current flowing through the resistor's material property.

These models and many other similar ones (e.g., Fourier's, Fick's, Hooke's) related to different engineering disciplines form the foundation of engineering, and it is through their application that we "trust" the behavior and responses of our "designs" in the real world. For a second, assume that you are riding an airplane that is designed based on laws and governing equations or models applied to fluid mechanics and solid mechanics,

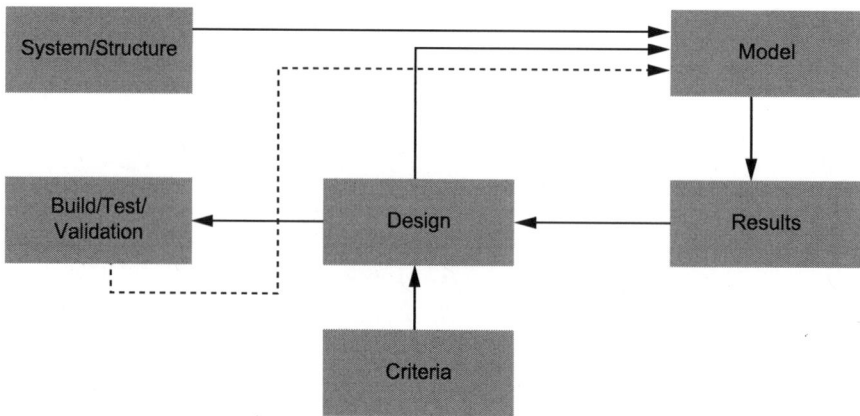

FIGURE 1.1 Modern design process for a system or component.

among others. If we don't trust and accept these laws and models, then it wouldn't be logical to ride an airplane!

Real-world phenomena are complex and usually involve many types of physics. For engineering application we usually simplify these phenomena and consider the dominant physics involved. For example, the length of a simple spring changes linearly under a given load according to Hooke's law. But it becomes a more complex problem if the spring's material behaves nonlinearly or if electrical charges flow through it. Traditionally the simplification of a problem is/was due to a lack of tools for finding a solution that could represent more accurately its "real-world" behavior. It is at this point that modern computational methods, such as FEM, and advanced modeling software tools, such as COMSOL, are valuable resources for finding solutions to complex engineering problems and optimizing our designs to have more economical, reliable, and durable products as end results.

Although the focus of this book is on using COMSOL as a modeling tool, we would like also to emphasize the importance and necessity of learning and hence understanding the foundation and mathematics behind FEM. For this purpose we cover a summary of the topic in Chapter 2 and provide references for further readings. We encourage readers to consult with available textbooks, online lectures and other trusted resources on this subject to enhance their technical and analytical backgrounds, which are key to becoming a competent modeler.

CHAPTER 2

FINITE ELEMENT METHOD—A SUMMARY

OVERVIEW

Finite element method (FEM) is the dominant computational method in engineering and applied science fields. Other methods including finite-volume, finite-difference, boundary element, and collocation are also used in practice. To provide general readers with a background for applications of FEM, either directly or with application of a software tool, we discuss the FEM principles in summary in this chapter. We also refer readers who are interested in further reading on this subject to a selection of available textbooks and references.

As discussed in Chapter 1, modeling has an ancient history. However, since the mid-twentieth century a new definition of modeling has gradually emerged (see References 2.1 and 2.2). This definition is a direct consequence of the development of advanced computational methods as well as huge advances in digital computers in terms of their CPUs and graphics processing power. As a result, computer modeling is synonymous to "modeling." The combination of advanced computational methods, applied mathematics, and powerful computers has created a valuable tool for engineers and applied scientists to model their designs/products before manufacturing them. The state-of-art modeling technologies that we currently enjoy using our laptop and mobile computers equipped with powerful software packages are the result of vast progress and advances in applied mathematics, computer science, engineering methods of analysis, and, of course, capital and business investment in these fields. For example, not long ago, just analyzing a tapered beam would take quite an amount of time and resources, whereas now an engineer can perform a similar—and accurate—analysis in about fifteen minutes!

As mentioned in the previous chapter, the mathematical models relevant to the physical phenomena involved in a given problem are the foundation of modeling. The mathematical model may be a system of algebraic equations, ordinary differential equations (ODE's), partial differential equations (PDE's), or more complex form of differential-integral equations in the form of a functional. Some of these equations are listed in Table 2.1 as examples where summation convention of indices is applicable. Among these models the ODEs and PDEs require application of suitable computational methods to find solutions for a set of given boundary conditions

$\sigma_{ij,j} + \kappa_i = 0$	Equilibrium equation (elliptic)
$\rho\left(\dfrac{\partial u_i}{\partial t} + u_j u_{i,j}\right) = p_{,i} + \mu u_{i,jj}$	Navier-Stokes equation (viscous fluid motion)
$\dfrac{\partial T}{\partial t} = \alpha\, T_{,jj}$	Diffusion-Heat equation (parabolic)
$\dfrac{\partial^2 \psi}{\partial t^2} = c\, \psi_{,jj}$	Wave equation (hyperbolic)
$F = k\Delta$	Hooke's law (algebraic)
$V = RI$	Ohm's law (algebraic)
$\nabla . E = 4\pi\rho$ $\nabla . B = 0$ $-\nabla \times E = \dfrac{1}{c}\dfrac{\partial B}{\partial t}$ $\nabla \times B = \dfrac{1}{c}\dfrac{\partial E}{\partial t} + \dfrac{4\pi}{c} J$	Maxwell's equations (electromagnetics)

TABLE 2.1 Examples of mathematical models for different phenomena.

for a given domain or geometry. FEM is one of these methods. It should be noted here that for some simple cases (e.g., simple geometries) exact solutions of the ODE/PDEs may exist, serving as valuable resources that can be used to validate the corresponding modeling results. However, in practice and especially for engineering applications, we use FEM to find solutions. This method becomes more valuable when a problem has complex geometry and/or complex boundary conditions.

FEM computation procedure starts with dividing the geometry of the problem at hand into several subdivisions. This step in the process is called *meshing*. As shown in Figure 2.1, the subdivisions or *elements* are simple geometrical shapes, such as triangle, quadrilateral, tetrahedron, or hexagon. Then the variation of the quantity in question that is governed by the relevant PDE/ODE (e.g., displacement, temperature, fluid pressure) is approximated using simple functions with the values of the dependent variables at the *nodes* of elements.

Using the calculus of variations to minimize a variational principle (see Reference 2.3) or a weighted-residual method, the governing PDE's are transformed into algebraic equations for each element. The equations obtained for each element are then collected to form the *global* system of algebraic equations that can be solved using available standard and advanced solvers. The solutions are the nodal values of the quantity in question (i.e., displacement, temperature, pressure). These nodal values can be used to calculate a quantity's values at any point inside each element and therefore over the whole geometrical domain of the problem.

Several advances have been made in the fields of computer-aided design (CAD), automatic meshing of complex geometries, robust solvers, elements formulations, post-processing, and overall "reliable" computational software packages. COMSOL is one of these packages that includes multiphysics modeling facilities. Recently, commercial CAD packages (e.g., Autodesk®, SolidWorks®) are equipped with some modeling facilities as well.

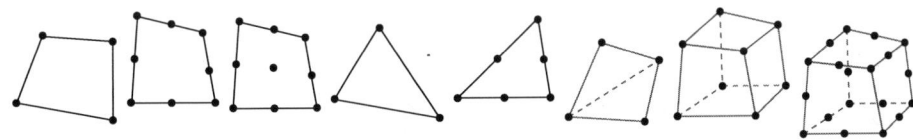

FIGURE 2.1 Examples of mesh *elements* and *nodes* (2D and 3D).

FEM FORMULATION

In general, there are two types of approaches to implement the FEM procedure described above:

1. Minimization of a functional using variational principle approach:

   In this approach, we use an integral expression instead of a PDE, and by using the calculus of variations we minimize this integral expression or the functional. It can be shown (see Reference 2.4) that this approach yields the same corresponding PDE or the governing equations (Euler-Lagrange equations) for a given problem. For example, the principle of minimum potential energy is a subset of this approach. This approach is used mostly in structural mechanics where we have the functional or can simply calculate it (e.g., using principle of virtual work). The matrix approach to finite element formulation can also be considered a subset of this approach.

2. Weighted residual approach:

   For many engineering problems either it is difficult or impossible to find an integral expression or functional that can be minimized and result in the governing or equilibrium equations. An example is the Navier-Stokes equations that govern viscous fluid flow. For these types of problems we use a weighted residual method and start the formulation directly from the corresponding PDE's. This method is mostly used for fluid flow, heat transfer, and nonlinear types of PDEs. See the chapter references to classical texts (e.g., Reference 2.5) for further readings on FEM.

Both of these methods are applicable for finite element formulation. The flowchart in Figure 2.2 depicts the process. We will use both of these methods (matrix and weighted residual) in the following examples to demonstrate the step-by-step FEM procedure accordingly. These simple

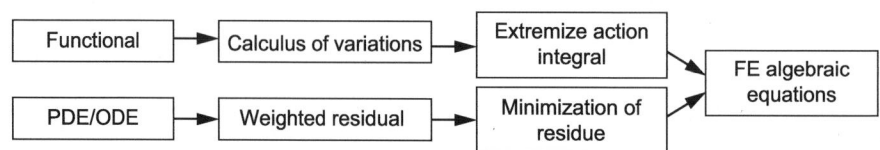

FIGURE 2.2 General process for finite element formulation.

examples are intended to serve the objectives of this chapter. More complex examples will be presented in future chapters.

MATRIX APPROACH

As mentioned above, the matrix approach can be considered a subset of the variational approach. For structures having members that can be considered as elements, such as trusses, the method is very straightforward. For analysis of a plate, which does not have obvious members or elements, we use the overarching variational approach. This method has commonly and traditionally been used for large structures such as aircrafts and tall buildings. This approach has its roots in the 1950s when pioneering engineers such as Turner et. al. (see Reference 2.6) and Clough (see Reference 2.7) developed and applied it to the analysis of aircraft structures.

Example 2.1: Analysis of a 2D Truss

For the 2D structure shown in Figure 2.3, calculate the displacement of the nodes/joints. Each node has 2 degrees of freedom (d.o.f.), which are displacements in x and y directions. All members of the truss have the same cross-section area A and modulus of elasticity E. Member 2-3 has a length of L, and vertical load P is applied at node 1. After finding the symbolic solution, numerical values can be assigned to the variables L, E, A, P for practical applications.

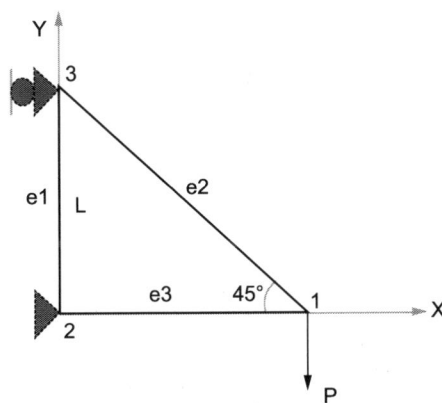

FIGURE 2.3 A 2D structure with truss elements.

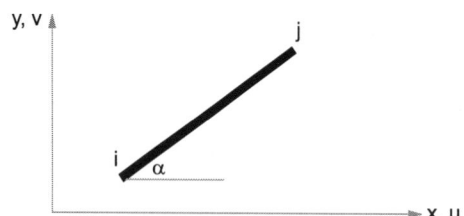

FIGURE 2.4 A general element with orientation with respect to x, y global coordinates.

Solution: This example can be solved using methods discussed in subjects like Statics and Strength of Materials. However, we would like to use this example for demonstrating the application of FEM using the matrix approach. The following steps explain this procedure:

First we assign node numbers (1, 2, 3) and element numbers (e1, e2, e3) to the nodes and members of this truss, as shown in Figure 2.3. The order is arbitrary; it will affect the resultant global matrix in general but not the final solution. Since each node has 2 d.o.f., each element/member has 4 d.o.f. and hence a 4 × 4 stiffness matrix. For reference, by definition the elements of stiffness matrix are forces per unit displacement for each node, and the direction of the force and displacement will determine the location of each element in the matrix. A general member *i-j* with orientation angle α is shown in Figure 2.4. The relationship for forces $F = (F_x, F_y)$ applied at each node (i or j) and the corresponding displacement $u = (u, v)$ can be written in a matrix format using the equilibrium equation $\{F\}=[K]\{u\}$, where $[K] = K_{ij}$ is the stiffness matrix of the truss members. These quantities are defined in the global system of coordinates (x, y).

For example, K_{11} is the force at node i = 1 in the x-direction due to a unit displacement at node i = 1 in the x-direction. Similarly, K_{12} is the force at node i = 1 in the x-direction due to a unit displacement at node 1 in the y-direction. Since [**K**] is a 4 × 4 matrix and truss elements experience only compression or tension load, then we need to transform these loads into the global system of coordinates (see Reference 2.4) to relate them to $\{F\}$ and $\{u\}$, as shown below:

$$\frac{AE}{L}\overbrace{\begin{bmatrix} c^2 & cs & -c^2 & -cs \\ cs & s^2 & -cs & -s^2 \\ -c^2 & -cs & c^2 & cs \\ -cs & -s^2 & cs & s^2 \end{bmatrix}}^{K_{ij}} \begin{Bmatrix} u_i \\ v_i \\ u_j \\ v_j \end{Bmatrix} = \begin{Bmatrix} F_{ix} \\ F_{iy} \\ F_{jx} \\ F_{jy} \end{Bmatrix}$$

Where $c = Cos(\alpha) = \dfrac{x_j - x_i}{L}$, $s = Sin(\alpha) = \dfrac{y_j - y_i}{L}$, $L = \sqrt{(x_j - x_i)^2 + (y_j - y_i)^2}$, and u_i is displacements at node i in x-direction and v_i is displacements at node i in y-direction. Using this formulation for each element/member of the truss, we can calculate their corresponding stiffness matrices. Note that angle α is measured counter-clockwise, or using the right-hand rotation rule for the system of coordinates shown in Figure 2.3.

- Member e1, node $i = 2$, node $j = 3$, $\alpha = 90°$, x2 = y2 = 0, x3 = 0, y3 = L; $s = \sin(90) = 1$, $c = \cos(90) = 0$; then;

$$k_{e1} = \dfrac{AE}{L}\begin{bmatrix} 0 & 0 & 0 & 0 \\ 0 & 1 & 0 & -1 \\ 0 & 0 & 0 & 0 \\ 0 & -1 & 0 & 1 \end{bmatrix}$$

- Member e2, node $i = 1$, node $j = 3$, $\alpha = (180 - 45) = 135°$, x1 = L, y1 = 0, x3 = 0, y3 = L; $s = \sin(135) = \sqrt{2}/2$, $c = \cos(135) = -\sqrt{2}/2$; then

$$k_{e2} = \dfrac{AE}{L\sqrt{2}}\begin{bmatrix} 0.5 & -0.5 & -0.5 & 0.5 \\ -0.5 & 0.5 & 0.5 & -0.5 \\ -0.5 & 0.5 & 0.5 & -0.5 \\ 0.5 & -0.5 & -0.5 & 0.5 \end{bmatrix}$$

- Member e3, node $i = 1$, node $j = 2$, $\alpha = 180°$, x1 = L, y1 = 0, x2 = y2 = 0; $s = \sin(0) = 0$, $c = \cos(180) = -1$; then

$$k_{e3} = \dfrac{AE}{L}\begin{bmatrix} 1 & 0 & -1 & 0 \\ 0 & 0 & 0 & 0 \\ -1 & 0 & 1 & 0 \\ 0 & 0 & 0 & 0 \end{bmatrix}$$

Now we need to combine the members' stiffness matrices, which are written in the global system of coordinates, to get the *global stiffness matrix* of the structure. For this we need to consider the degrees of freedom at each node (i.e., 2) and the total number of nodes (i.e., 3). Hence the global stiffness matrix would be 6 × 6 since (2 × 3 = 6). We write a 6 × 6 matrix and fill its elements as follows:

- Element 1 has the nodes 2 and 3, so its stiffness matrix will occupy rows and columns (2i − 1 and 2i); 2 × 2 − 1 = <u>3</u> and 2 × 2 = <u>4</u> and 2 × 3 − 1 = <u>5</u> and 2 × 3 = <u>6</u>

    In row 3, columns 3, 4, 5, and 6 put (0, 0, 0, 0)

    In row 4, columns 3, 4, 5, and 6 put (0, 1, 0, −1)

    In row 5, columns 3, 4, 5, and 6 put (0, 0, 0, 0)

    In row 6, columns 3, 4, 5, and 6 put (0, −1, 0, 1)

- Element 2 has the nodes 1 and 3, so its stiffness matrix will occupy rows and columns (2i − 1 and 2i); 2 × 1 − 1 = <u>1</u> and 2 × 1 = <u>2</u> and 2 × 3 − 1 = <u>5</u> and 2 × 3 = <u>6</u>

    In row 1, columns 1, 2, 5, and 6 put (1, −1, −1, 1) / 2√2

    In row 2, columns 1, 2, 5, and 6 put (−1, 1, 1, −1) / 2√2

    In row 5, columns 1, 2, 5, and 6 put (−1, 1, 1, −1) / 2√2

    In row 6, columns 1, 2, 5, and 6 put (1, −1, −1, 1) / 2√2

- Element 3 has the nodes 1 and 2, so its stiffness matrix will occupy rows and columns (2i − 1 and 2i); 2 × 1 − 1 = 1 and 2 × 1 = 2 and 2 × 2 − 1 = 3 and 2 × 2 = 4

    In row 1, columns 1, 2, 3, and 4 put (1, 0, −1, 0)

    In row 2, columns 1, 2, 3, and 4 put (0, 0, 0, 0)

    In row 3, columns 1, 2, 3, and 4 put (−1, 0, 1, 0)

    In row 4, columns 1, 2, 3, and 4 put (0, 0, 0, 0)

The resulting global stiffness matrix is:

$$K_{global} = \frac{AE}{L} \begin{bmatrix} 1/2\sqrt{2}+1 & -1/2\sqrt{2} & -1 & 0 & -1/2\sqrt{2} & 1/2\sqrt{2} \\ -1/2\sqrt{2} & 1/2\sqrt{2} & 0 & 0 & 1/2\sqrt{2} & -1/2\sqrt{2} \\ -1 & 0 & 1 & 0 & 0 & 0 \\ 0 & 0 & 0 & 1 & 0 & -1 \\ -1/2\sqrt{2} & 1/2\sqrt{2} & 0 & 0 & 1/2\sqrt{2} & -1/2\sqrt{2} \\ 1/2\sqrt{2} & -1/2\sqrt{2} & 0 & -1 & -1/2\sqrt{2} & 1/2\sqrt{2} \end{bmatrix}$$

Now we apply the boundary conditions and known forces applied on the truss. These are $u_2 = v_2 = u_3 = 0$. The final system of algebraic equations is:

$$\frac{AE}{L}\begin{bmatrix} 1/2\sqrt{2}+1 & -1/2\sqrt{2} & -1 & 0 & -1/2\sqrt{2} & 1/2\sqrt{2} \\ -1/2\sqrt{2} & 1/2\sqrt{2} & 0 & 0 & 1/2\sqrt{2} & -1/2\sqrt{2} \\ -1 & 0 & 1 & 0 & 0 & 0 \\ 0 & 0 & 0 & 1 & 0 & -1 \\ -1/2\sqrt{2} & 1/2\sqrt{2} & 0 & 0 & 1/2\sqrt{2} & -1/2\sqrt{2} \\ 1/2\sqrt{2} & -1/2\sqrt{2} & 0 & -1 & -1/2\sqrt{2} & 1/2\sqrt{2} \end{bmatrix} \begin{Bmatrix} u_1 \\ v_1 \\ 0 \\ 0 \\ 0 \\ v_3 \end{Bmatrix} = \begin{Bmatrix} 0 \\ -p \\ R_{2x} \\ R_{2y} \\ R_{3x} \\ 0 \end{Bmatrix}$$

R's are reaction forces at the supports, as shown in Figure 2.5. We have as a result, six unknowns (u_1, v_1, v_3, R_{2x}, R_{2y}, R_{3x}) and six equations that can be solved for the unknowns.

The solution of Example 2.1 is complete here. We should also note that the stiffness matrix method or displacement method (as compared to flexibility matrix method or force method) is very suitable for computer programing. However, we need to obtain the reduced form of the global stiffness matrix to avoid the singular matrix, which means to avoid rigid-body motion. This task is performed by implementing the boundary conditions.

General Procedure for Global Matrix Assembly

When the number of elements for a structure is large, combining the stiffness matrices of the elements becomes a difficult task sometimes prone to error. To help with this task and get the right global stiffness matrix, we present a formula that can be used for any number of nodes and d.o.f. in a structure.

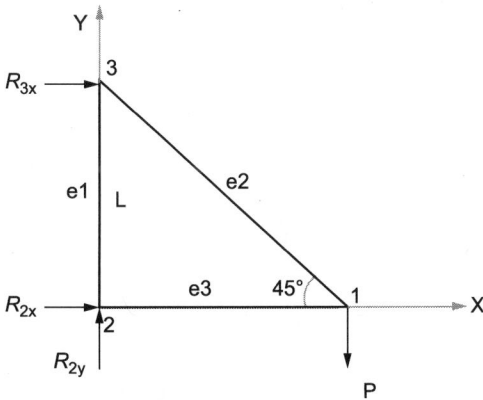

FIGURE 2.5 Free-body diagram, reaction forces, applied load.

Let's assume that we have n number of d.o.f. (n can take a value from 1 to 6) for each node. Then for a triangular element, which has 3 nodes, we have a $3n \times 3n$ stiffness matrix. The values of stiffness matrix components of this triangular element should find their right places in the global stiffness matrix as follows (say for node number j):

$$nj-(n-k)$$

where k is an integer; $k=1, 2, \ldots, n$. Similarly for an element with 4 nodes, such as a quadrilateral, we have a $4n \times 4n$ stiffness matrix, for which the above equations can be used. We demonstrate the application of this useful formula in the next example.

Example 2.2: Global Matrix for Triangular Elements

The elements of the stiffness matrix for the triangular shape element are given as $[k] = k_{ij}$, as shown in Figure 2.6. This element has 3 nodes (1, 3, 5). Calculate the location of each element of $[k]$ in the global stiffness matrix when the total number of nodes is 5 (i.e., the whole structure has 5 nodes) and the d.o.f. for each node is 3.

Solution: Since each triangular element has 3 nodes, then its element stiffness matrix is 9×9;

$$[k] = \begin{bmatrix} k_{11} & k_{12} & \cdots & k_{19} \\ k_{21} & k_{22} & \cdots & k_{29} \\ \vdots & \vdots & \vdots & \vdots \\ k_{91} & k_{92} & \cdots & k_{99} \end{bmatrix}; k_{ij} = k_{ji}$$

Here we assume that we have the numerical values of k_{ij}. Using the formula $nj-(n-k)$, for ($n = 3$, $k = 1, 2, 3$) we have:

for j = 1; {3 × 1 − (3 − 1) = **1**, 3 × 1 − (3 − 2) = **2**, 3 × 1 − (3 − 3) = **3**}

for j = 3; {3 × 3 − (3 − 1) = **7**, 3 × 3 − (3 − 2) = **8**, 3 × 3 − (3 − 3) = **9**}

for j = 5; {3 × 5 − (3 − 1) = **13**, 3 × 5 − (3 − 2) = **14**, 3 × 5 − (3 − 3) = **15**}

Since the global stiffness matrix is 15×15 ($3 \times 5 = 15$), the above elements would take the following places in the global stiffness matrix: {1, 2, 3, 7, 8, 9, 13, 14, 15}. The procedure to place these elements in the global stiffness matrix is as follows: We place the elements of row 1 from k_{ij} (k_{11}, k_{12}, ... k_{19}) in row 1 and columns (1, 2, 3, 7, 8, 9, 13, 14, 15) of the global matrix. Similarly, elements of row 2 of k_{ij} are placed in row 2 and columns (1, 2, 3,

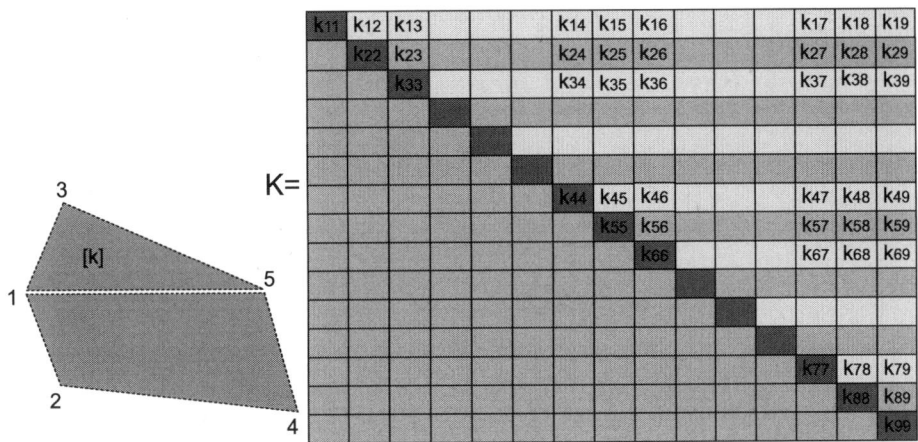

FIGURE 2.6 A triangular shape element and global stiffness matrix.

7, 8, 9, 13, 14, 15) of the global matrix. We continue in the same manner for rows 3, 7, 8, 9, 13, 14, and 15. The resulting global stiffness matrix **K,** with elements of k_{ij}, and using symmetry, is shown in Figure 2.6.

WEIGHTED RESIDUAL APPROACH

As mentioned in previous section, for any PDE, in principle, there exists an integral equation that can be minimized using variational principle, and the resulting Euler-Lagrange equation will be the same as the original PDE. Minimization of the corresponding integral is, usually but not always, easier than solving the original PDE. However, finding this integral equation is not always easy or for some equations not yet obtained (e.g., Navier-Stokes). It can be shown (see Reference 2.8) that the method of weighted residual is an equivalent approach that can be applied to any PDE and is also more suitable for computer programming. There are several weighted residual methods; among them we will discuss the Galerkin method, which is the most commonly used FEM in engineering.

GALERKIN METHOD

Assume that a PDE is represented by $L(\Phi) = 0$, where Φ is a dependent variable and in general can be a function of space and time, and L is the differential operator (e.g., ∇^2, ∇). Now, if we take the approximate solution of the function Φ to be $\tilde{\Phi}$, then $L(\tilde{\Phi}) \neq 0$, obviously. Let's assume the error to be R, then we have:

$$L(\tilde{\Phi}) = R$$

By using a weighted residual method we minimize this error over the entire domain in which the PDE applies, and therefore the approximated solution asymptotes to the exact one. The integral of R over the computational domain V using a weight function W can be written as:

$$\int_V RW \, dV = 0$$

For the Galerkin method, W is the same as *shape functions*. We will explain the concept and applications of shape functions in Example 2.3, but interested readers can refer to finite element texts (see Reference 2.4). For Finite-Volume method, $W=1$; for Collocation method, $W=\delta(x)$; and for spectral method, $W=$Fourier series. By dividing the whole domain into several elements and applying the weighted residual method for each element, we end up with a system of algebraic equations. This system then can be solved using known boundary and initial values for the dependent variable Φ. It should be mentioned here that the guessed function $\tilde{\Phi}$ should satisfy the initial and boundary conditions, as well. The Galerkin method should include the essential (or kinematic constraints) and will result in the nonessential (or natural dynamic) ones.

SHAPE FUNCTIONS

As mentioned previously and demonstrated in Example 2.3, obviously the choice of shape functions is an important step in the FE analysis. The shape function can be piecewise linear, quadratic, or cubic polynomials. More complex functions are also considered for different types of elements. We can use higher order polynomials for the shape functions and/or increase the number of elements to reach an acceptable accuracy level and convergence for a finite element analysis.

In finite element software packages, this choice is referred to as the "element type" for modeling.

In this section, we go a bit deeper to explain the concept and application of shape functions.

As mentioned previously in FE, we use piecewise functions (usually polynomials) for approximating the variation of the dependent variables inside elements. Let's assume a linear function as $u = a_1 + a_2 x$, which defines a linear variation for u over the domain of an element with length L. Also let u_1 and u_2 be the values of u at nodes or end points of the element. Then we have

$$u\big|_{x=0} = u_1 = a_1$$

$$u\big|_{x=L} = u_2 = u_1 + a_2 L \Rightarrow a_2 = \frac{u_2 - u_1}{L}$$

$$\therefore \quad u = u_1(1 - \frac{x}{L}) + u_2 \frac{x}{L}$$

As it should, this function satisfies the boundary conditions. We define $N_1 = 1 - \frac{x}{L}$, $N_2 = \frac{x}{L}$ as shape function for function u. Then we can write the function u as

$$u = N_1 u_1 + N_2 u_2 \quad or$$

$$u = [N_1 \quad N_2] \begin{Bmatrix} u_1 \\ u_2 \end{Bmatrix}$$

Note that the shape function $N_1 = 1$ and $N_2 = 0$ at $x = 0$ and $N_2 = 1$ and $N_1 = 0$ at $x = L$. This is a very important property of shape functions. In general for multiple elements, we can write

$$u = \sum_{i=1}^{n} u_i N_i$$

where n is the total number of nodes of the elements.

CONVERGENCE AND STABILITY

The solution of an FE model should asymptote/approach to the corresponding exact solution, with "acceptable" accuracy. In other words, the results should converge towards the "exact" solution in a "stable" manner. The stability of the solution indicates that the error should not oscillate in a way that the answer becomes infinite. We want the answer of the FE model approaches exact solution in a monotonic or "damped" oscillation manner. There are two ways to achieve the convergence:

- **h-type** convergence, or by increasing the number of elements or the resolution of the mesh; and

- **p-type** convergence, or by increasing the element interpolation/shape function order, which employs higher order polynomials for shape functions.

Usually p-type approach provides a faster convergence, for a given problem, towards exact solution, but in practice h-type is used more often because computer time and power are readily available.

Example 2.3: Heat Transfer in a Slender Steel Bar

A slender steel bar with length L is given, as shown in Figure 2.7. At $x = 0$ the bar is subjected to heat flux q and at the other end is kept at a fixed temperature T_L. All other surfaces are thermally insulated. Calculate the temperature distribution along the bar using FEM.

Solution: The temperature T satisfies the following differential equation and boundary conditions (steady state, 1D):

$$-k\frac{d^2T}{dx^2} = 0, T = T_L \ @ \ x = L \text{ and } -k\frac{dT}{dx} = q \ @ \ x = 0$$

Note that this problem has an analytical solution that can be easily found by integrating the corresponding ODE twice, or $T = T_L + \frac{q}{k}(L-x)$. The exact solution, if available, is useful for checking the FE results. However, in this example we want to use FEM to find the solution. This is done by following the steps below:

- We divide the steel bar into two sections, or *elements* (e1 and e2), hence we will have 3 nodes, $i = 1, 2, 3$, as shown in Figure 2.8.

- Within each element, we approximate the temperature distribution by using a simple function, as: $\tilde{T}(x) = \sum_{i=1}^{3} N_i(x)T_i = N_1 T_1 + N_2 T_2 + N_3 T_3$

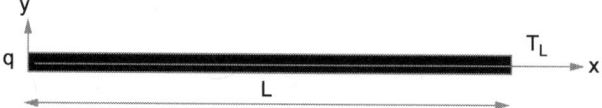

FIGURE 2.7 A slender steel bar with heat flux and temperature boundary conditions.

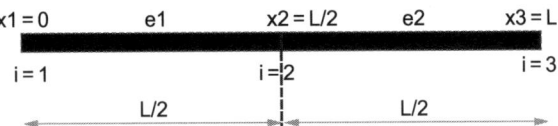

FIGURE 2.8 Elements and nodes numbers for the steel bar.

N_i is the linear shape function that defines the temperature distribution within element e_i and T_i is the value of temperature at node i. The approximated function $\tilde{T}(x)$ should satisfy the boundary conditions, always. The shape function N_i should be equal to 1 at node i and zero at the other node of corresponding element.

Now we introduce the $\tilde{T}(x)$ into the governing equation, which results in a residue since the approximate function will not satisfy the PDE, hence $-k\dfrac{d^2\tilde{T}}{dx^2} = R \neq 0$. We cannot force the residue to vanish at any location on the steel bar, but we can have its integral (or its total net value) vanish. To do this we use the method of minimization of weighted residuals, specifically Galerkin method, as mentioned previously. Therefore we will have

$$\int W\left[-k\frac{d^2\tilde{T}}{dx^2}\right]dx = \int N(x)R\,dx = 0$$

Note that in Galerkin method shape function $N(x)$, is used as the weight function.

We integrate the above equation, using the method of integration by parts

$$\int N[-k\frac{d^2\tilde{T}}{dx^2}]dx = \int [k\frac{d\tilde{T}}{dx}\frac{dN}{dx}dx] - kN\frac{d\tilde{T}}{dx}\bigg|_0^L = 0$$

This is called *weak* formulation for FEM, since the second-order derivative is transformed into first-order one (hence a "weaker" constraint on the variation of the independent variable, in this case the temperature $\tilde{T}(x)$). Now we substitute for $\tilde{T}(x) = \sum_{i=1}^{n+1} N_i(x)T_i = N_1T_1 + N_2T_2 + N_3T_3$ then

$$\int_0^{L/2}[k\frac{dN_i}{dx}(\sum_{j=1}^{3}\frac{dN_j}{dx}\tilde{T}_j)]dx + [N_i(-k\frac{d\tilde{T}}{dx})]_0^{L/2}$$
$$+ \int_{L/2}^{L}[k\frac{dN_i}{dx}(\sum_{j=1}^{3}\frac{dN_j}{dx}\tilde{T}_j)]dx + [N_i(-k\frac{d\tilde{T}}{dx})]_{L/2}^{L} = 0 \quad for\ i = 1,2,3$$

Using the shape functions definition, we have as a result a system of 3 equations for 3 unknown values $\tilde{T}_1, \tilde{T}_2, \tilde{T}_3$. The final equations in matrix form are

$$\begin{bmatrix} 2k/L & -2k/L & 0 \\ -2k/L & 4k/L & -2k/L \\ 0 & -2k/L & 2k/L \end{bmatrix} \begin{bmatrix} \tilde{T}_1 \\ \tilde{T}_2 \\ \tilde{T}_3 \end{bmatrix} = \begin{bmatrix} q \\ 0 \\ -q_L \end{bmatrix}$$

But $\tilde{T}_3 = T_L$, therefore the third equation can be solved separately. Finally, we have

$$\begin{bmatrix} 2k/L & -2k/L \\ -2k/L & 4k/L \end{bmatrix} \begin{bmatrix} \tilde{T}_1 \\ \tilde{T}_2 \end{bmatrix} = \begin{bmatrix} q \\ 0 \end{bmatrix} + \frac{2k}{L} T_L \begin{bmatrix} 0 \\ 1 \end{bmatrix}$$

Or $\tilde{T}_1 = qL/k + T_L$ and $\tilde{T}_2 = qL/2k + T_L$.

We see that these are actually equal to the values of exact solutions at the nodes $i = 1, 2, 3$ since the exact variation of temperature in the rod is a linear one, similar to our assumed shape functions. The solution of this example is complete here.

This example demonstrates an application of Galerkin method. Readers may want to review and pay attention to the process of FEM, weak formulation of FE, and weighted residual minimization among other concepts presented in this example. For more examples, readers may refer to Reference 2.3.

EXERCISE PROBLEMS

Problem 2.1: Solve Example 2.1 for numerical values A = 10 cm^2, E = 200GPa, L = 1.5 m, P = 200 N.

Problem 2.2: Repeat Problem 2.1 for α = 53°.

Problem 2.3: Using the method described in Example 2.2, write down the global stiffness matrix for the 3-element structure as shown below. [**K**], [**M**], and [**L**] are elements' stiffness matrices.

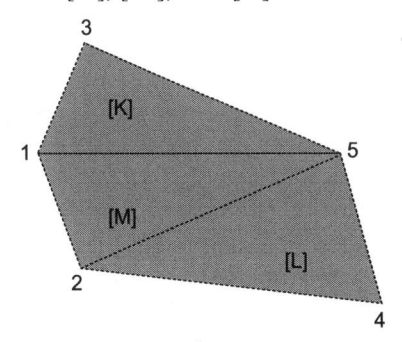

Problem 2.4: Renumber the nodes of the plate structure shown in Figure 2.6 and repeat the assembly of global stiffness matrix. Compare your results with those of Problem 2.3.

Problem 2.5: Solve Example 2.3 with a set of new boundary condition given as; T = T$_1$ at x = 0, and T = T$_2$ at x = L.

REFERENCES

2.1: Clough, R.W., "Original Formulation of the Finite Element Method," *J. Finite Element. Anal. Des.* 7, no. 2, 1990.

2.2: Robinson, J., "Early FEM Pioneers," Robinson and Associates, Dorset, England, 1985.

2.3: Shames, Irving H. and Dym, Clive L., *Energy and Finite Element Methods in Structural Mechanics*, Hemisphere Publishing Corp, 1985.

2.4: Cook, Robert D., *Concepts and Applications of Finite Element Analysis*, 2nd ed., John Wiley & Sons, Inc., 1981.

2.5: Baker, A. J., *Finite Element Computational Fluid Mechanics*, McGraw-Hill, 1983.

2.6: Turner, M., Clough, R., Martin, H., and Topp, L., "Stiffness and deflection analysis of complex structures," *J. Aero., Sci.* 23, no. 9, 1956, pp. 805–823.

2.7: Clough, R.W., "The Finite Element Method in plane stress analysis," Proc. 2d Conf. Electronic Computation, American Society of Civil Engineers, Pittsburg, PA, 1960, pp. 345–378.

2.8: Pilkey, Walter D. and Wunderlinch, Walter, *Mechanics of Structures: Variational and Computational Methods*, CRC Press, 1994.

CHAPTER 3

COMSOL—A MODELING TOOL FOR ENGINEERS

OVERVIEW

In this chapter, we introduce COMSOL and its features as a software tool for modeling. The objective is to provide a "tour" of this software package and introduce its features, modules, and facilities to readers as well as provide guidelines for building models using COMSOL. To demonstrate COMSOL module applications, we will provide several modeling examples and their applications in detail in the next chapter. Because it would be exhaustive to include all features available in COMSOL in a single book, our main objective is to provide a collection of examples and modeling guidelines through which readers could build their own models.

COMSOL, which is a finite-element based modeling tool, has a well-developed graphic user interface and several modules for modeling common and advanced types of physics involved in engineering and applied science practices. Its history goes back to when this package was called FEMLAB™ and was written based on MATLAB™ whereas newer versions are stand-alone packages. The latest version is COMSOL 4.x series. In version 4, the user interface was upgraded and more modules and stronger, smarter solvers were added. Meshing with COMSOL is almost seamless and "automatic," yet it gives users the choice of having custom-designed mesh both for structured and unstructured types. It has a rich materials property database and yet allows users to define their own database for their desired materials.

Another major feature of COMSOL is the ability to solve any PDE/ODE that users might have and which may not fit into classical governing equations (e.g., wave, heat, equilibrium). A new feature in version 4, called (0D), allows users to solve problems that do not have space as a relevant defined dimension, such as electric or thermal equivalent networks. The most recent feature to become available allows COMSOL to be run directly through a CAD software package interface, such as SolidWorks™ and some Autodesk™ products.

The author's experience with this package includes its ongoing improvement in features and especially the solvers, which makes it an efficient modeling tool for small- to medium-size problems (in terms of geometrical size) and with multiphysics involved. In COMSOL, users can see the governing equations for the type of physics they solve right on the interface—a feature that is very helpful for assigning right values to the variables and boundary conditions as well as knowing what type of equations you are solving using the finite element method. Building the geometry of a model is possible either by using CAD facilities available in COMSOL or using live communication modules such as, LiveLink™. LiveLink modules are available for major CAD packages (e.g., Inventor™, SolidWorks™, SpaceClaim™) as well as MATLAB™ and Excel®. The post-processing features allow users to study and see the modeling results and analyze them using color-coded surface graphs and data line graphs, among others. The Reports feature is very useful, enabling users to generate a file using modeling results in common word processing formats.

COMSOL has comprehensive and rich Help documentation as well as tutorials. In the following sections, we will take a tour of COMSOL features and modules and introduce some of its features. COMSOL provides free workshop and webinars for new users as well as more extensive training courses for a fee. More information about the features, models gallery, and tutorials is available on the COMSOL Website (*www.comsol.com*).

COMSOL INTERFACE[2]

After purchasing the product license you can install COMSOL on your machine, either PC or Mac. When launching COMSOL, a window similar

---

[2] Related to version 4.3. new version 4.4 has a different interface.

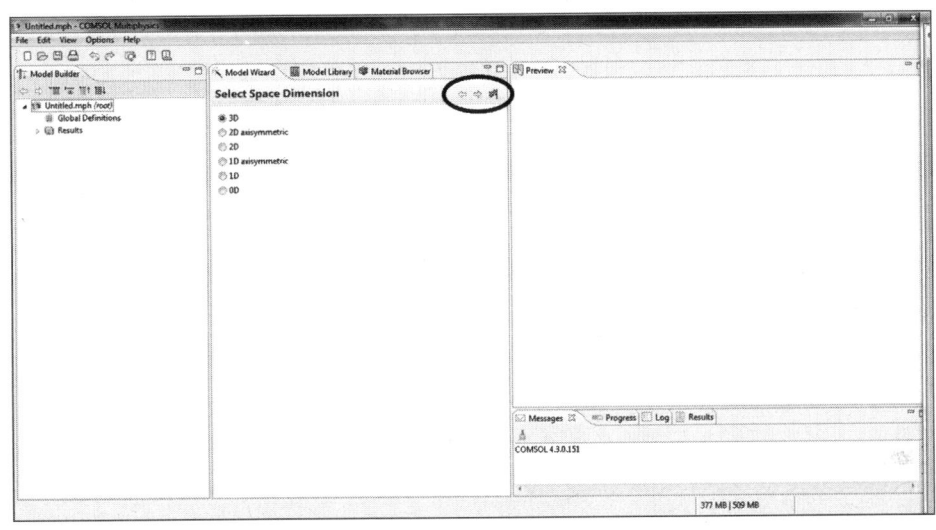

FIGURE 3.1 Entrance window; opens when launching COMSOL 4.3.

to the one shown in Figure 3.1 will open (default layout windows arrangement may vary). This window has three major parts:

- Model Builder, which accumulates all features/steps that may be used for building a model in a tree-style format;
- Model Wizard, which allows data entry for applications of different features, such as geometry, physics, material properties, meshes, and post-processing; and
- Graphics, which shows geometry as well as the modeling results in color with legends.

Each of these main windows has tabs that allow users to change the content of the corresponding window, if needed, and to move them, detach them from the main window, or change the layout.

A user starts building a model by operating in the Wizard window. The sequence of actions is listed and recorded in the Model Builder tree, which becomes handy for editing and maneuvering around the model. The graphics window will show the results in different stages during construction from geometry to final modeling results.

Through Model Wizard facilities, a user can choose the geometrical dimension of the model (0D to 3D) including axisymmetric 1D or 2D

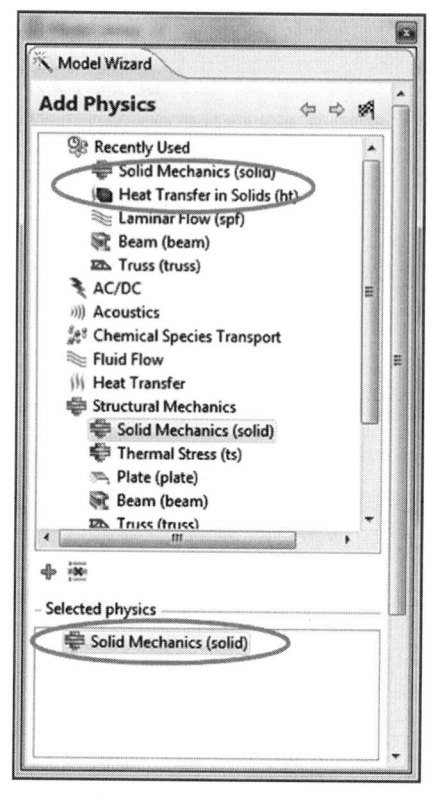

FIGURE 3.2 Add Physics feature window for COMOSL 4.3.

problems. Once a dimension is chosen, by pressing the arrow key (➡, marked by the ellipse in Figure 3.1) a user can move to the next window to choose the right physics for the model, as shown in Figure 3.2. Available physics modules depend on the software license purchased; additional features can be added to the basic license. The user can select the required physics for the problem by right-clicking the options from the Model Wizard list. The selections are then listed in the *Selected physics* space at the bottom of the window, as shown in Figure 3.2.

Clicking the arrow on the right-top corner will take the user to next window to choose the Study Type. Several studies are available based on the features in the purchased license as well as the physics chosen. See Figure 3.3 for Solid Mechanics physics as an example. The desired study type can be chosen by clicking on it and then clicking on the flag icon located on the top right corner of the window (see Figure 3.3)

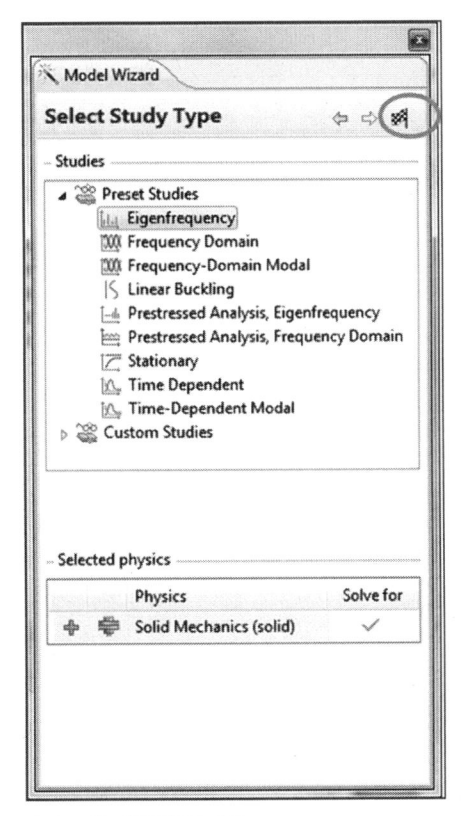

FIGURE 3.3 Select Study Type window for COMSOL 4.3.

At this stage the Model Wizard window is replaced by the Geometry window, as shown in Figure 3.4, and is ready for drawing the model's geometry. Drawing facilities are shown on the top of the window, in the icon bar.

Building a model geometry can be done in two ways in COMSOL: by using the drawing tools to build relatively simple geometry inside COMSOL, or by importing geometry from a commercial CAD package[3]. The main module for seamlessly importing a geometry file is called *LiveLink*™ *Interfaces* and should be purchased as part of the user license. When a user has a solid model of a geometry open in a CAD package, through LiveLink it is possible to seamlessly import the file as well as communicate

---

[3] Inventor, SolidWorks, Pro/Engineer, SpaceClaim, Creo Parametric. Please check COMSOL.com for updates.

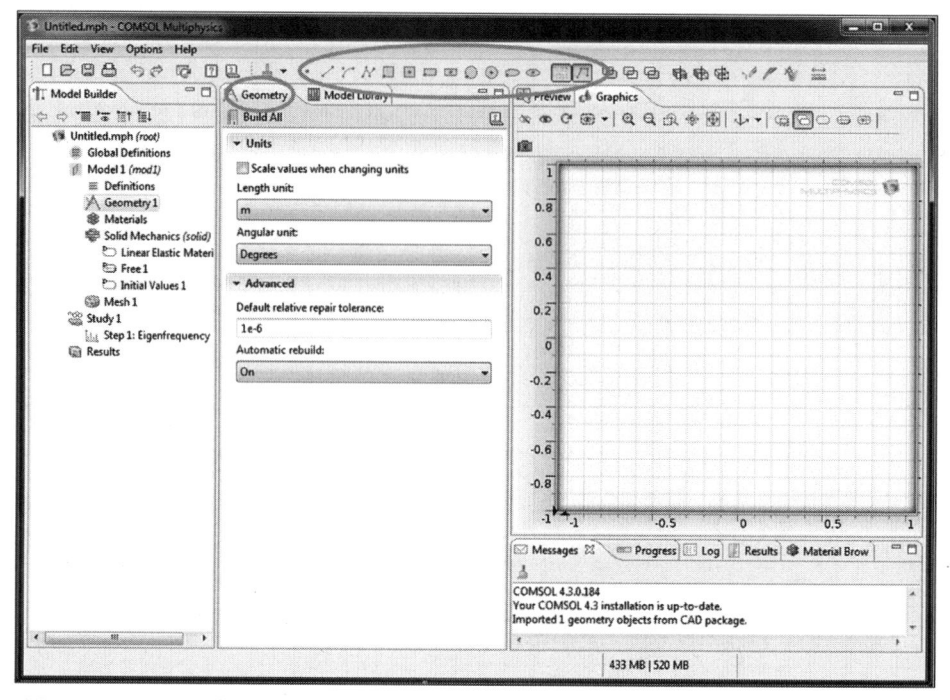

FIGURE 3.4 Geometry building window in COMSOL 4.3.

between the parameters of the CAD geometry in COMSOL and update it if needed. The LiveLink modules make the CAD package an integral part of COMSOL. These modules enable users to run COMSOL from the interface of a CAD package.

LiveLink modules are extremely useful and are recommended for users who have more complex model geometry and would like to use their favorite CAD package for building them. In addition, users can import a solid model in conventional format such as Parasolid, STEP, IGES, VRML, and STL. See the COMSOL manuals for CAD import modules.

Several Unit Systems of measurement are available in COMSOL. Users can choose their desired Units by clicking on the (*root*) at the top of Model Builder window, as shown in Figure 3.5, and scrolling down in the Root tab to find the Unit System. When a Unit System (e.g., SI) is chosen at the *root* level, users can enter the data in different units but COMSOL automatically converts it to the main selected Unit System.

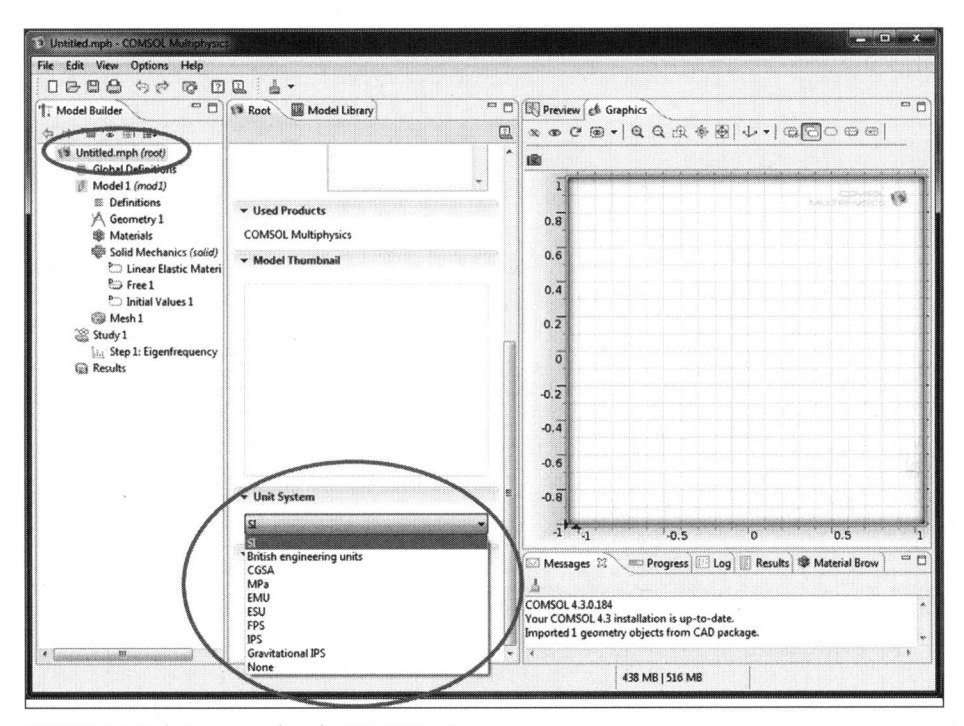

FIGURE 3.5 Unit System options in COMSOL 4.3.

By clicking on the Geometry 1 node in the Model Builder window, the Geometry window will open again. After building the geometry (or importing it), material properties can be added to the geometry by either entering the data directly or using the database available in COMSOL. This is done by right-clicking on the Materials node in the Model Builder window and choosing the Material Browser, as shown in Figure 3.6. As shown, materials are categorized for different applications and users also can add their own material library to the list.

Boundary and Initial conditions can be selected by right-clicking on the selected/corresponding physics in the Model Builder window. When selected, the conditions can be implemented to the boundaries by clicking the corresponding edges or surfaces of the model geometry in the Graphics window. A typical boundary selection is shown in Figure 3.7.

Meshing, a process for dividing the geometry into finite elements, is a major step in modeling. Meshing facility in COMSOL is generally done

30 • COMSOL FOR ENGINEERS

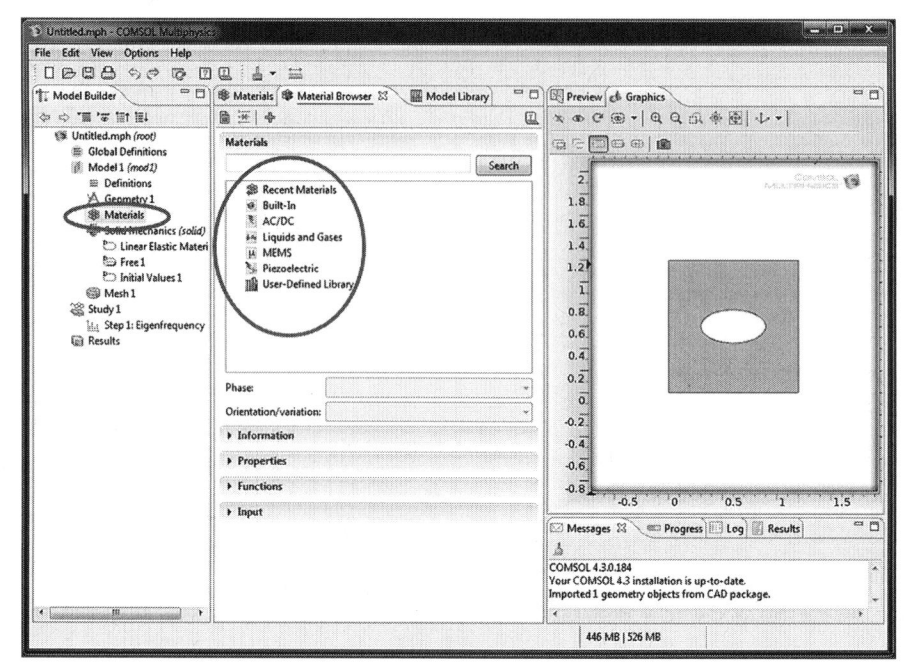

FIGURE 3.6 Material Browser and Library in COMSOL 4.3.

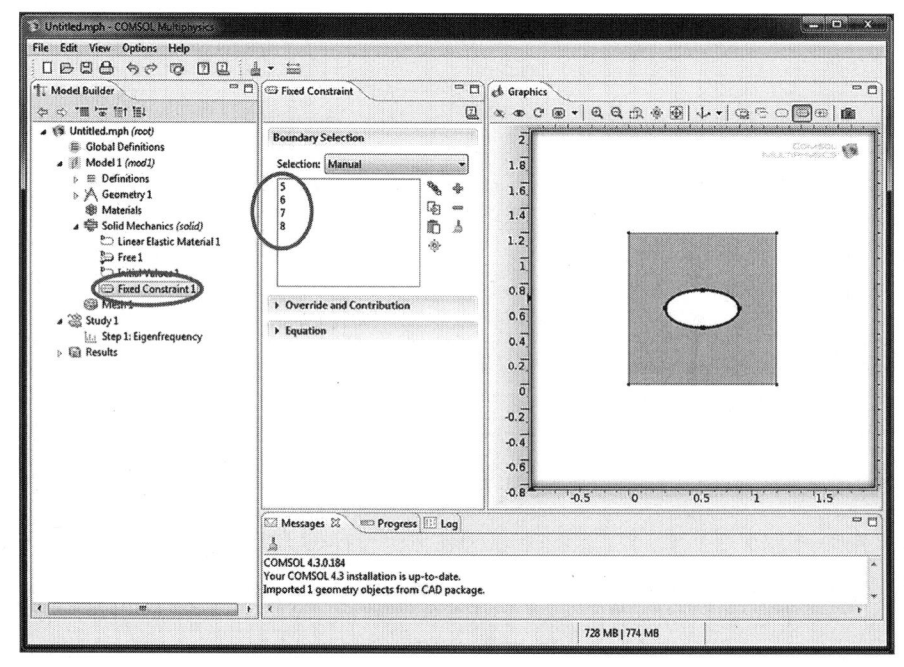

FIGURE 3.7 Boundary condition selection in COMSOL 4.3.

using *free* meshing or *mapped* mesh. Free meshing is unstructured mesh, and mapped meshing is similar to structured mesh. Users have options to modify the resolution of the mesh for different regions of the geometry. Mixed or hybrid meshing is also possible by combining structured and unstructured mesh types. Users can also define the resolution or number of meshes/elements in different regions of the geometry. COMSOL has adaptive-mesh facility, a feature that customizes the mesh resolution during the solution process for complex geometry or physics. Meshing facility is not shown at the start window; this will appear after building or importing a geometry. Clicking on the *Mesh 1* node in the Model Builder opens the Mesh window, as shown in Figure 3.8. Users can choose the resolution of the mesh and also manipulate the meshing parameters in this window by choosing options from Mesh Settings.

At this point the model is ready for analysis. This is done by right-clicking on the Study node in the Model Builder window and choosing Compute (or ▬). Additional studies can be implemented by right-clicking on the (*root*) point in the Model Builder/tree and choosing Add Study.

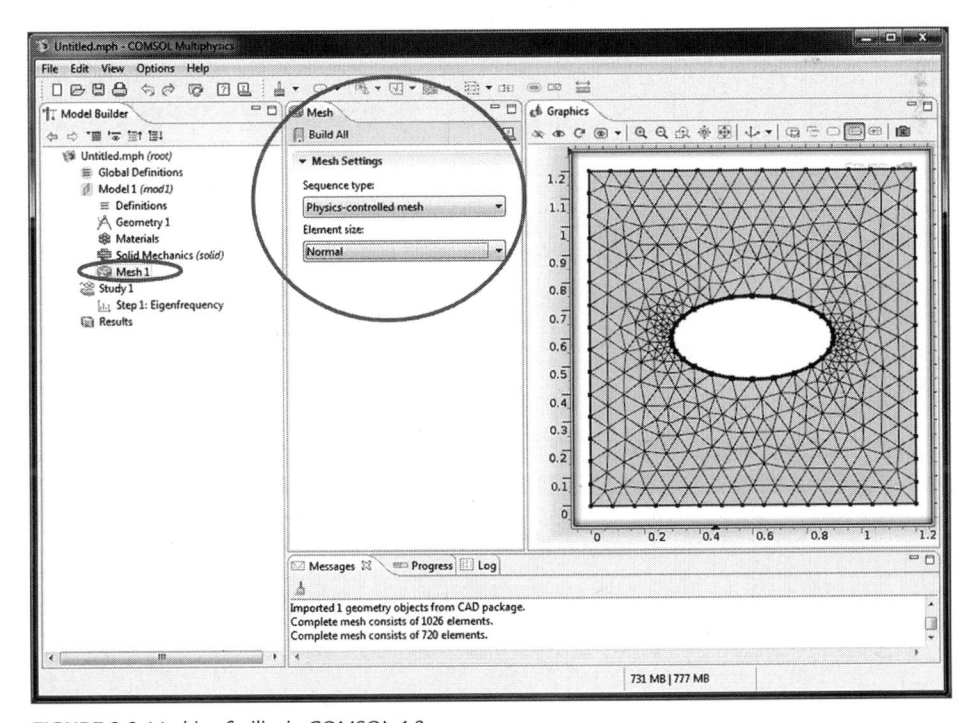

FIGURE 3.8 Meshing facility in COMSOL 4.3.

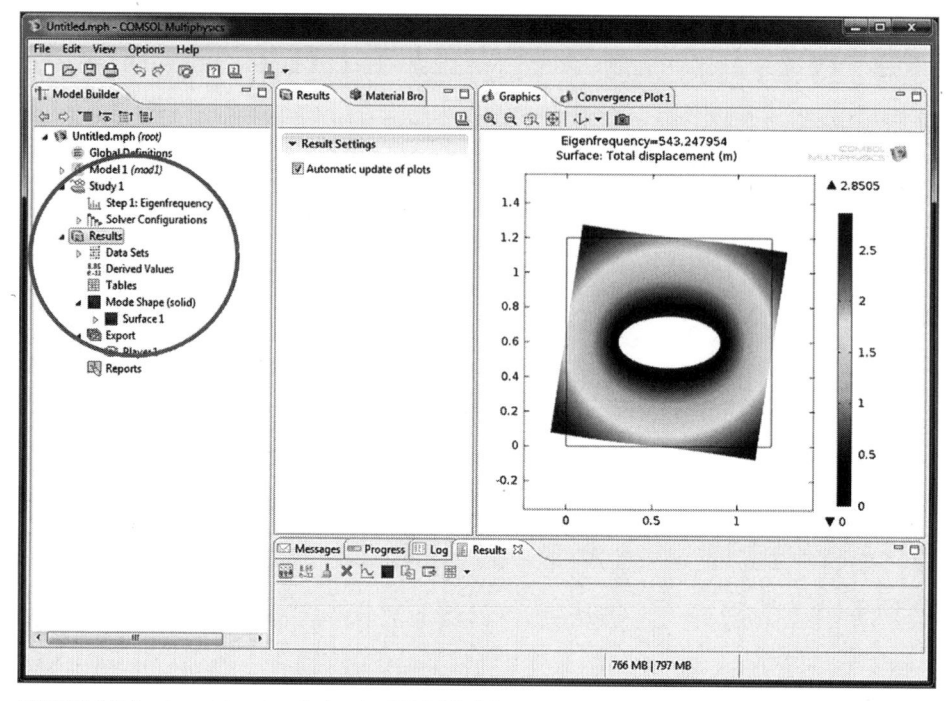

FIGURE 3.9 Study and Results window in COMSOL 4.3.

After running the simulation for the selected Study, the results are displayed in the Graphics window, as shown in Figure 3.9.

A detailed or summary report can be generated by clicking on the Reports node in the Model Builder window. This is a very useful feature for communicating and filing model results.

COMSOL MODULES

COMSOL has many ready-to-use modules to handle modeling most, if not all commonly occurring engineering problems. In addition, users can solve unconventional governing equations/PDEs using available Mathematics modules in COMSOL.

Following is a list of COMSOL physics/application modules available for purchase. Additional features and modules are released with newer versions of the software. For an updated and complete list, check the COMSOL Website (*www.comsol.com*).

- CAD Import Module
- CFD Module
- Structural Mechanics Module
- Heat Transfer Module
- Optimization Module
- AC/DC Module
- Mathematics Module
- Chemical Transport Module
- Acoustics Module
- Batteries & Fuel Cells Module
- Geomechanics Module
- MEMS Module
- RF Module
- Subsurface Flow Module
- Particle Tracing Module
- Pipe Flow Module

COMSOL MODEL LIBRARY AND TUTORIALS

In addition to standard technical support, COMSOL offers several resources including free workshops and webinars to assist new users or provide updated information for new versions as well as extended workshop sessions available for a fee.

When a user installs COMSOL, many other resources become available to support their modeling tasks. One resource is the Model Library, which offers solved models for training, teaching, or modification. Registered users can download these models and supporting documents (usually in PDF format). Models available in the library are a valuable way to start a model with similar or closely related physics and modify them according to a specific modeling problem. A sample list of models in the Model Library is shown in Figure 3.10, for a typical sublist for CFD Module. The Model Library is available under the View button in the toolbar, as shown.

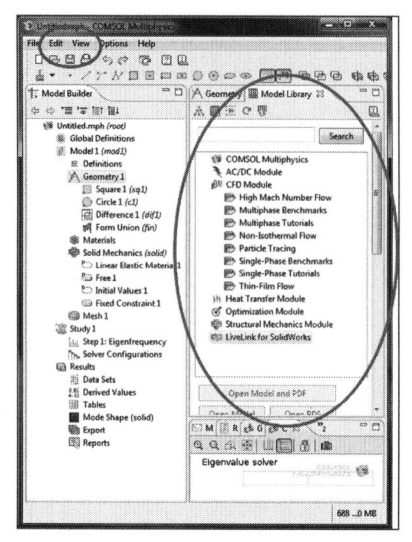

FIGURE 3.10 Sample list of models in Model Library, COMSOL 4.3.

There are also two types of Help documents available for users under the Help button in the toolbar: Documentation and Dynamic Help. The Help Documentation offers users access an extensive, searchable list of documents that explain interface icons and keys, as well as details of modules, physics, meshing, geometry, post-processing, and more. Users at varying levels of expertise can refer to the documentation to find more details about COMSOL features as well as answers to specific questions. The Dynamic Help feature opens a specific section of the Help Documents relevant to the section or feature in use at the moment. This is helpful feature for finding assistance when users want to find out about the specifics of, say, fixed constraints 1 boundary condition, as shown in Figure 3.11.

GENERAL GUIDELINES FOR BUILDING A MODEL

The major sequential steps for building a model using COMSOL for a given problem are as follows:

1. Define the problem, including physics and materials involved.

2. Identify the governing equations and boundary conditions to have a clear understanding of the scope of the problem's solution.

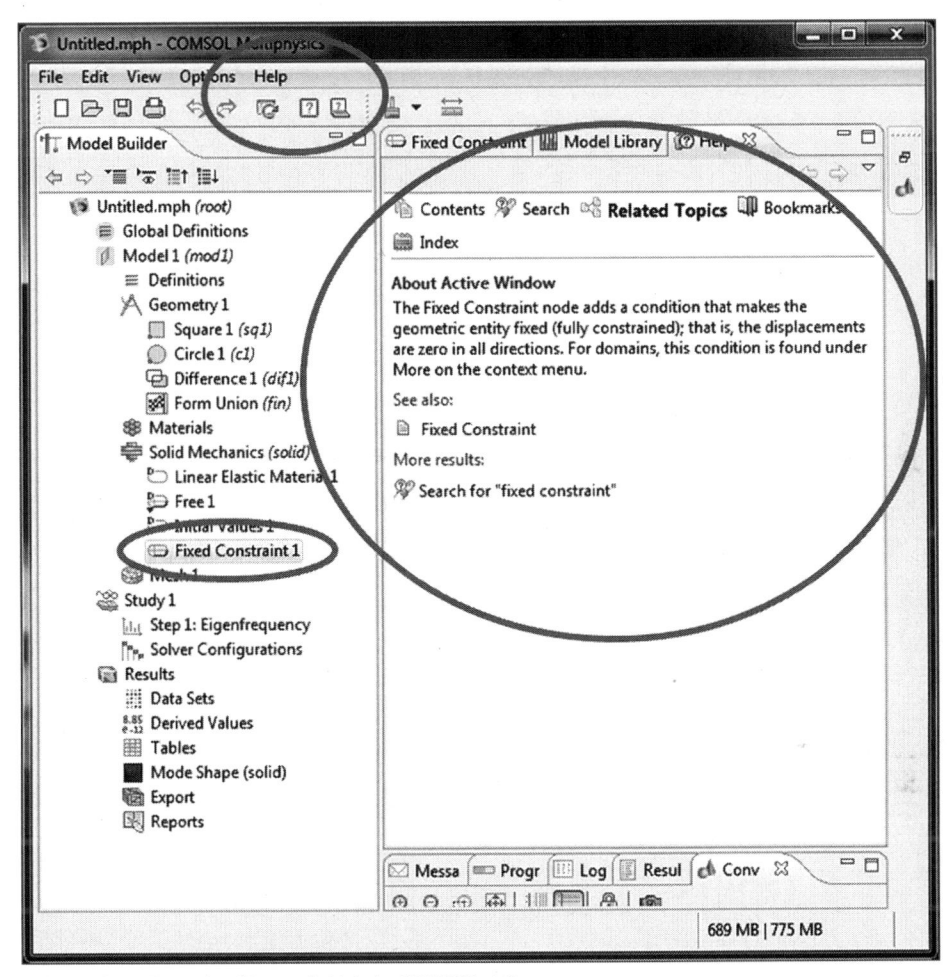

FIGURE 3.11 Example of Dynamic Help in COMSOL 4.3.

3. Launch COMSOL.

4. Use COMSOL features to assign the dimension (1D, 2D, 3D, etc.), the physics involved, and the temporal (steady or transient, etc.) of the problem.

5. Build the geometry of the problem (if required), import your CAD file, or use LiveLink to access your model geometry.

6. Assign material properties to the built geometry blocks of the problem.

7. Add physics according to Steps 1 and 2.
8. Create a mesh or finite elements for the built geometry.
9. Solve/run the model.
10. Visualize the results and validate them, either using hand calculations or comparing to similar known results.
11. Create a report of the model and its specifications.

CHAPTER 4

COMSOL MODELS FOR PHYSICAL SYSTEMS

OVERVIEW

Many engineering problems involve either one type of physics or, if they involve multiphysics, can be simplified to a dominant one. For example, consider the laminar flow of a fluid at a high value of Reynolds number. Since a Reynolds number may be interpreted (see Reference 4.1) as the ratio of inertia over viscous forces exerted on the fluid, then for a fluid with a high value Reynolds number we can neglect viscous forces and consider the fluid as an ideal frictionless one. When turbulence effects are present, flow becomes more complicated and equivalent turbulence-induced viscous stresses (so called Reynolds stress) should be included. Or in the case of a steel beam, for example, we can neglect the effects of deformation due to shear stresses at the cross-section of the beam and assume that the beam cross-section remains perpendicular to the axis of the beam about which it is bent. These types of approximations in engineering are very common and sometimes are matters of "engineering art" or technical judgment. It takes much experience to simplify a problem without losing the dominant physics and obtain results that are useful in practice and have applications.

In this section, we use COMSOL to model example problems covering a wide range of concepts including stationary equilibrium, dynamic equilibrium, buckling, time-dependent, fluid flow, heat transfer, electrical circuits, and transport phenomena. The main objective is to provide, for users and readers, a collection of solved examples that could be applied directly or lead to further solutions of similar or more complex problems using COMSOL. It is assumed that readers are familiar with relevant engineering principles and governing equations. Nevertheless, in each example we offer brief explanation of physics involved along with governing equations and phenomena, as applicable. It is recommended that readers cover Chapters 1 through 3 before attempting examples in this chapter. As mentioned previously, the examples are solved using version 4.3. Newer version (like 4.4) could also be used to solve these examples with finding the relevant tools in the new COMSOL 4.4 interface. Readers could also open the files from the accompanied CD in version 4.4 and study or modify the examples pertinent to their modeling needs.

SECTION 4.1: STATIC AND DYNAMIC ANALYSIS OF STRUCTURES

In this section, we use COMSOL features to model some problems in structural mechanics. Models include static, dynamic, parametric study, and buckling of two- and three-dimensional structures.

Example 4.1: Stress Analysis for a Thin Plate Under Stationary Loads

A plate is a structural member that can carry loads acting normal to or along its plane. The resulting stresses due to bending, normal and shear forces applied cause deformation. For thin plate the deformation due to shear force along the thickness is ignored, whereas for thick plate it is, usually, taken into account. A plate is considered thin when its thickness over a typical horizontal dimension of the plate is smaller than or equal to 0.1. Thick plate theory is that of Mindlin's and Reissner's (see Reference 4.2). Governing equations are the equilibrium equations, which are available in the COMSOL manual. We will refer to these equations in the solution steps.

As an example we consider a thin plate (12 in. by 8 in.), as shown in the Figure 4.1. We would like to calculate the displacement and stress field in the plate resulting from the applied loads. Plate thickness is 0.4 in, and applied tension stress load is 500 psi at the right-middle edge. Modulus of elasticity of the plate material is 3E7 psi, and Poisson's ratio is 0.3. The boundaries on the left side of the plate are constrained while the other plate sides are let free.

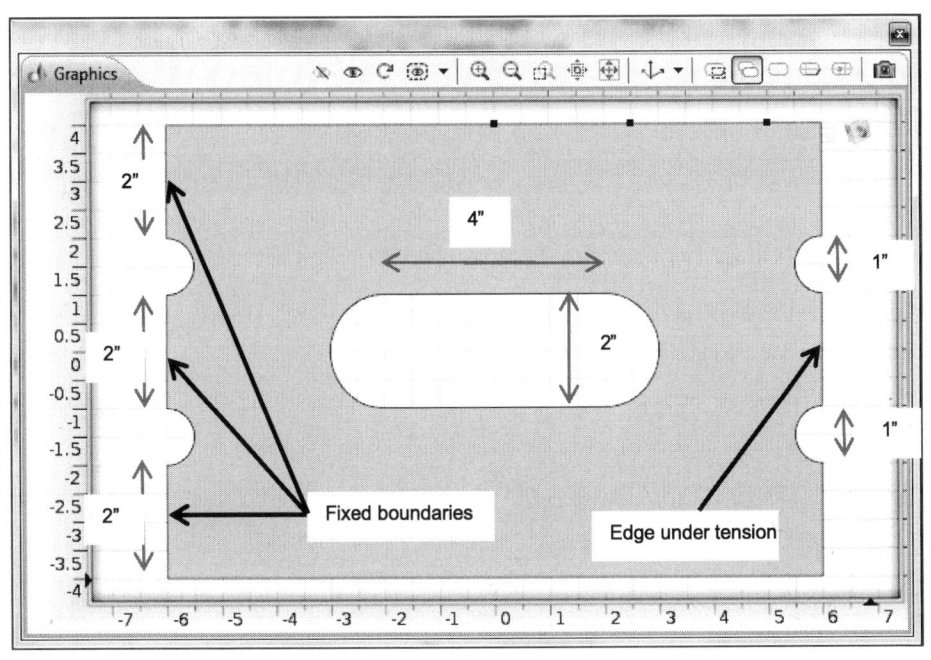

FIGURE 4.1 Geometry, dimensions, and boundary conditions.

Solution

1. Open COMSOL and open a new file (either by clicking on the new button icon, □, clicking on File in the toolbar and choosing New, or right-clicking on the (*root*) node in the Model Builder and choosing Add Model).

2. Save the new file as Example 4.1 by clicking on File>Save as in the toolbar.

3. Choose the geometry dimension by clicking on 2D in the Model Wizard, and then click the arrow icon (⇨) to go to the next step, as shown in Figure 4.2.

4. Choose the Physics by moving the pointer to the Structural Mechanics node (located in the Model Wizard window). Open the list by clicking on the pointer triangle located on its left, choose Plate by clicking on it, and then click on the plus icon ✚ to add selected physics). Alternatively, you can right-click on Plate in the list to add the selected

physics. Plate should be listed in the selected physics, as shown in Figure 4.3. Also notice that the dependent variables are listed, as displacements (u, v, w), at the bottom of the Model Wizard window. Next, click on the arrow icon to move to the next step.

5. Select the type of analysis or study by clicking on Stationary (in the Study list) and clicking on the flag icon to finish the problem setup. You should see that the Geometry window opens.

6. In the Geometry window, open the Length unit section and choose (in) for inches, the dimensions of this problem. (Notice that the default Unit System used in COMSOL is SI. In this example, we keep it "as is" knowing that the software automatically does the unit conversion.)

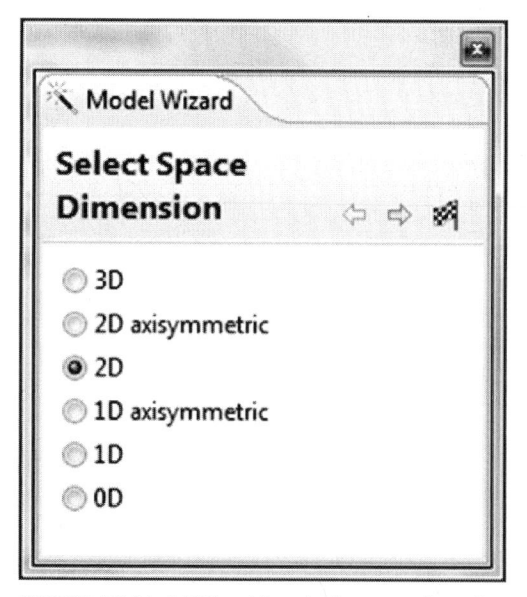

FIGURE 4.2 Model Wizard for selecting space dimension.

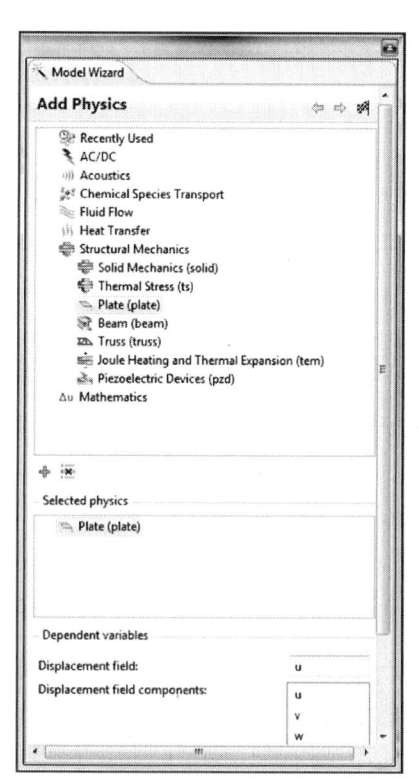

FIGURE 4.3 Model Wizard for selecting and adding physics.

7. In the Model Builder window, click on the Geometry 1 node, shown in Figure 4.4 to open a list of drawing tools in the toolbar. Draw the geometry of the plate as follows:

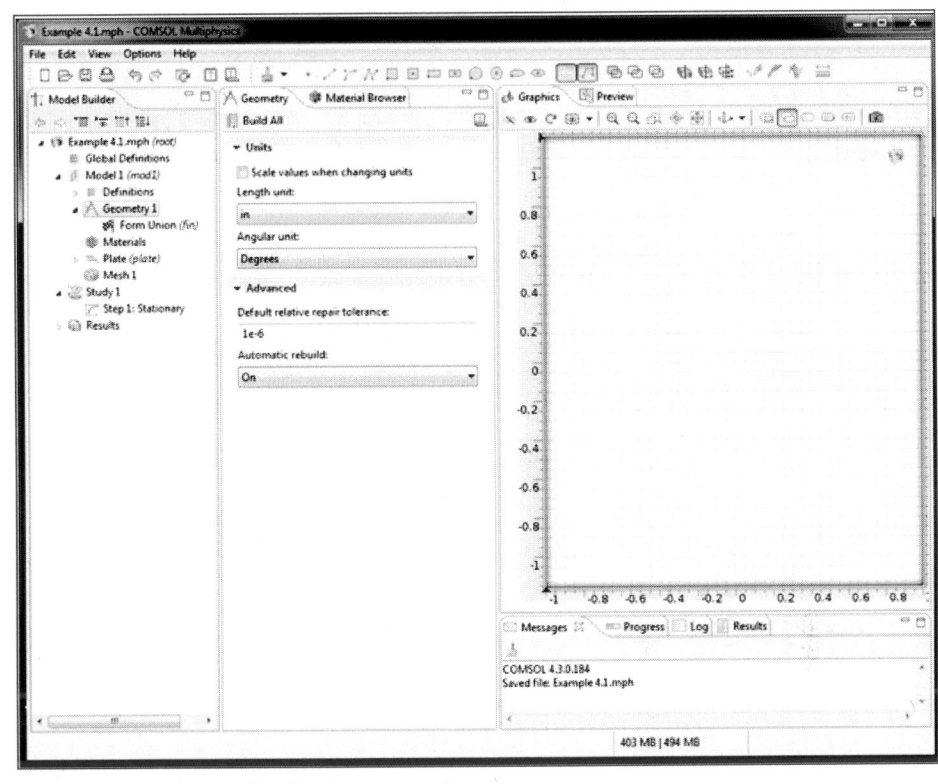

FIGURE 4.4 Graphics window for Geometry entries.

   7.1 Click on the rectangle icon (Draw Rectangle (Center)) on the toolbar and draw a rectangle in the Graphics window. The Rectangle 1 will be created in the Model Builder under the Geometry 1 node. Click on the rectangle, and then in the Rectangle window that opens enter Width = 12 in., Height = 8 in., and Position for Center is (0, 0). Then click on the Build Selected icon located at the top of the Rectangle window. Click on the Zoom Extents icon to scale the Graphics window to display the entire rectangle, as shown in Figure 4.5.

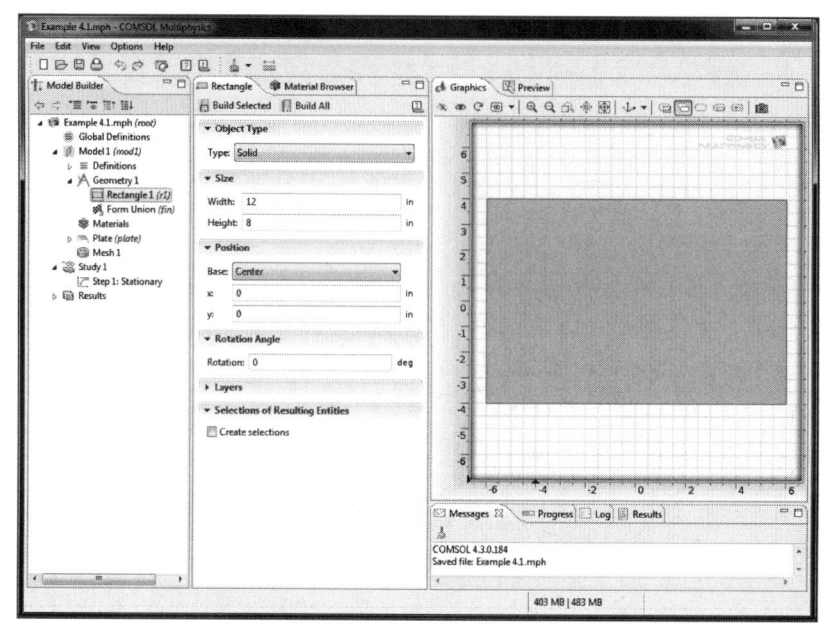

FIGURE 4.5 Graphics window with a rectangle.

> **7.2** Draw four circles using the circle drawing tool, each with radius of 0.5 in. and their centers (Circle 1: x = –6, y = 1.5 in.), (Circle 2: x = –6, y = –1.5 in.), (Circle 3: x = 6, y = 1.5 in.), (Circle 4: x = 6, y = –1.5 in.). Occasionally you may have to click on the Geometry 1 node in the Model Builder tree to see the drawing tools. Results are shown in Figure 4.6.
>
> **7.3** Click at any empty location in the Graphics window, and then press the CTRL key and A to choose all objects in the geometry window. The color of all objects will change to red, indicating that they are selected. Then click on the Geometry node in the Model Builder window and click on the Difference icon located in the toolbar. By pressing the Difference icon ⊟, a Boolean operation is performed that subtracts the circles from the rectangle. Results are shown in Figure 4.7.
>
> **7.4** Similarly, draw a rectangle (Width = 4 in., Height = 2 in., Center at x = 0, y = 0) and subtract it from the resulting plate geometry. Again, draw two more circles with radii 1 in. (Circle 5: x = –2, y = 0 in.), (Circle 6: x = 2, y = 0 in.) and subtract them to create the resulting plate geometry shown in Figure 4.1.

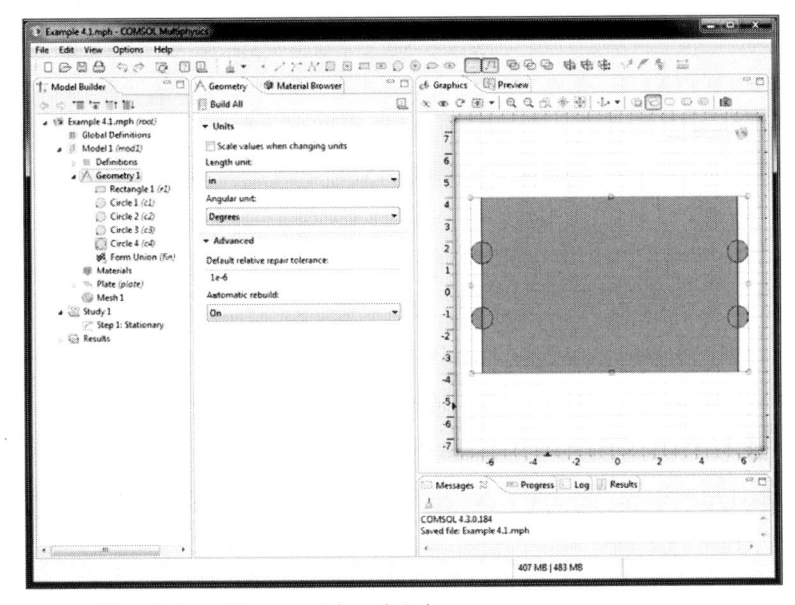

FIGURE 4.6 Graphics window with a rectangle and circles.

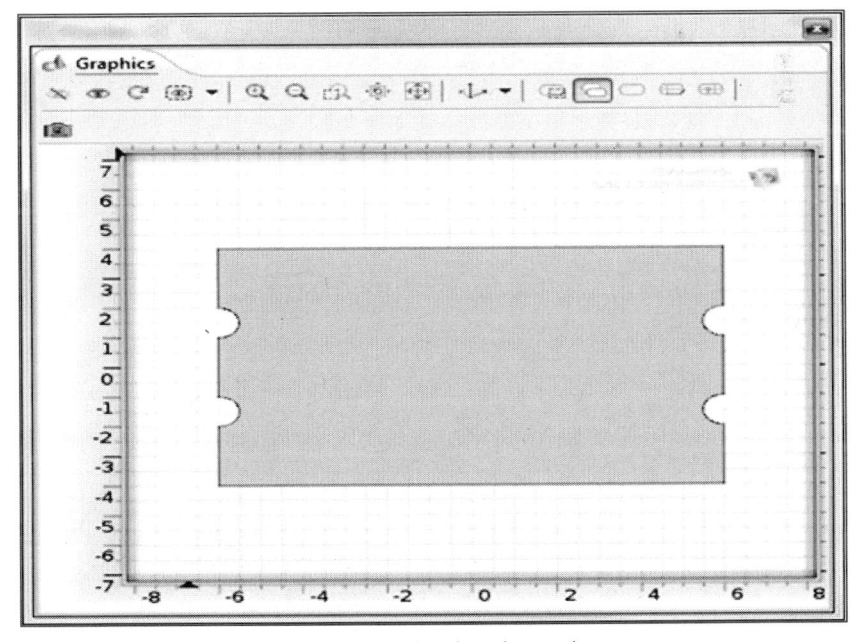

FIGURE 4.7 Graphics window with a rectangle and circles subtracted.

7.5 Click on the Plate node in the Model Builder window, and then open the Thickness section and enter 0.4[in] to assign the thickness of the plate. Notice the default is the SI unit (i.e., meters). It is optional but recommended to open the list under the Equation section to see the governing equations for the plate to be solved using finite element method.

8. The next step is assigning the material properties. Move the pointer to the Plate node in the Model Builder window and open the list by clicking the triangle icon. Click on the Linear Elastic Material 1 node to open the Linear Elastic window. Scroll down to see the Linear Elastic Material section. Under Young's modulus select User defined and enter 3e7[lb/in^2], and under Poisson's ratio select User defined and enter 0.3, as shown in Figure 4.8. Notice that the default SI unit (Pa) is set for Young's modulus. COMSOL automatically converts the data from psi to Pa.

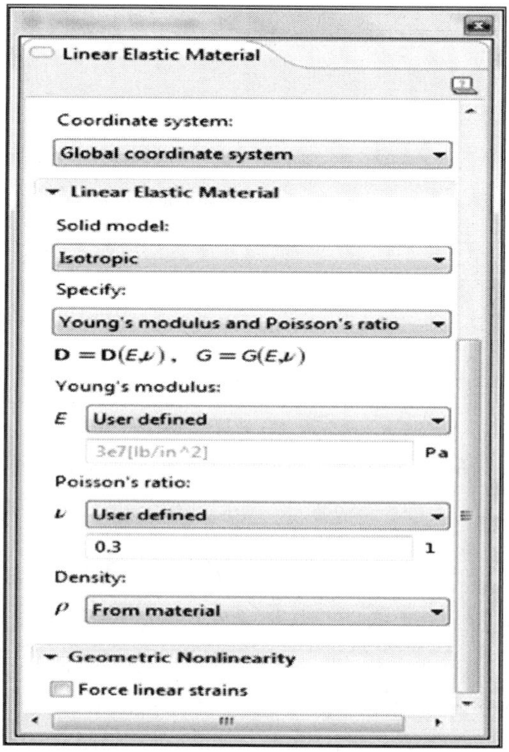

FIGURE 4.8 Data entry window for Linear Elastic Material.

9. Assign the boundary conditions by right-clicking the Plate node in the Model Builder window and choosing Fixed Constraint. Click on the Fixed Constraint 1 node to open the corresponding window. In the Graphics window, click on the left edges of the plate and add them to the Boundary Selection list by clicking on the plus icon in the Fixed Constraint window. (Alternatively, you can click on the edges of the plate and then right-click to add them to the Boundary Selection list). Results are shown in Figure 4.9.

10. Set the applied load by right-clicking on the Plate node in the Model Builder window and choosing Edge Load. The Edge Load 1 will be created under the Plate node. Click on the Edge Load 1 and choose the middle-right side edge of the plate in the Graphics window, and add it to the Boundary selection list by clicking on the plus sign icon. Then choose the Load defined as force per unit area in the Force section (shown in the Edge Load window) and enter 1500[lb/in^2] in the x-component of the Edge load, since the load is applied as a tension on the edge of the plate. Results are shown in Figure 4.10.

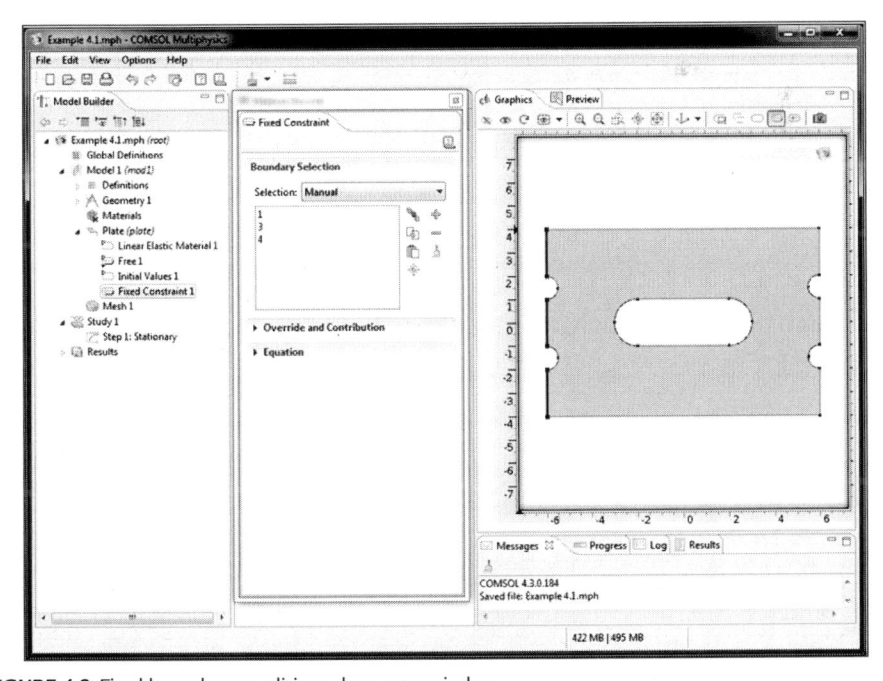

FIGURE 4.9 Fixed boundary conditions data entry window.

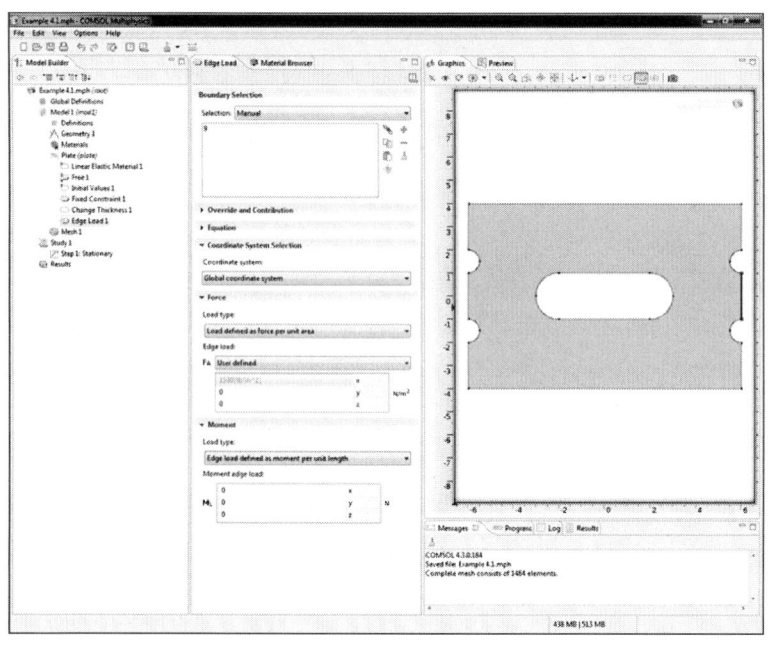

FIGURE 4.10 Load boundary conditions data entry window.

11. Create a mesh by clicking on the Mesh node in the Model Builder window, and in the Mesh settings section choose Physics-controlled mesh for Sequence type and Normal for Element size. Then click Build All to create the mesh. Statistics for the mesh are shown in the Messages window (usually located under/close to the Graphics window). A mesh with 1464 elements for this example was created). Results are shown in Figure 4.11.

12. To run the model, right-click on the Study 1 node in the Model Builder window and select Compute ≡. The COMSOL default result for von Mises stresses is shown in the Graphics window. For this model, it took 11 seconds to run the finite element model with total 18,516 d.o.f. (as registered in the Messages window) on a typical laptop computer.

13. To manipulate the results, click on Surface 1 node under the Stress Top (*plate*) node in the Model Builder window. This will open the Surface window. In this window, open the list under Unit in the Expression section and choose psi, then open the Title section, choose Manual from the Title type, and enter von Mises stress for Example 4.1 (psi). Click on the (x-y) icon located on the Graphics

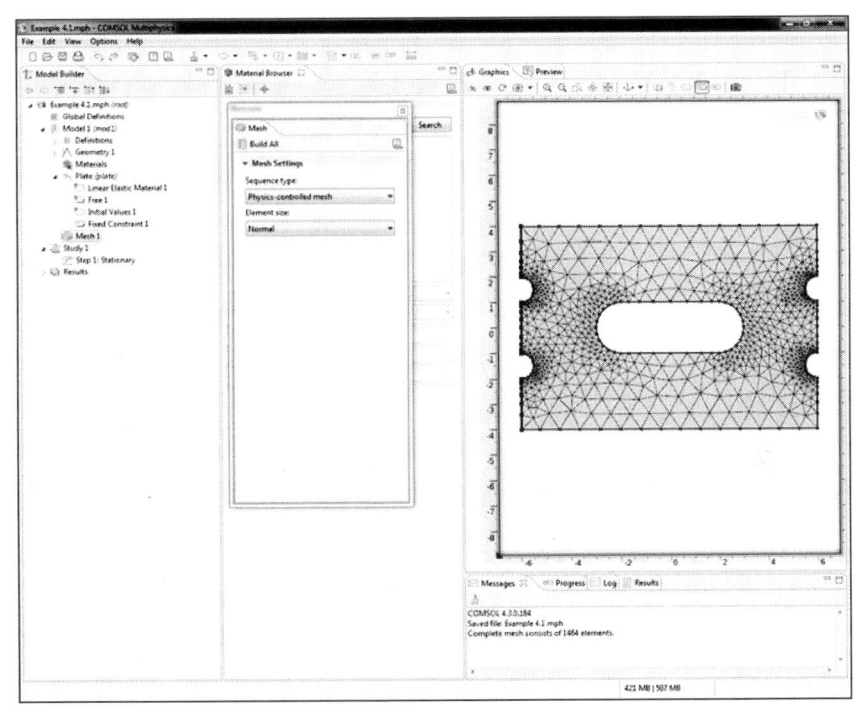

FIGURE 4.11 Mesh data entry window and resulting mesh.

window toolbar once to set the view. Then click on the plot icon. Occasionally, you may need to zoom-unzoom the view to fit the results in the Graphics window. To show the displacement results, click on the Replace Expression icon (located on the right side of the Expression section) and choose Plate>Displacement>Total displacement (*plate.disp*). Change the Title type to Automatic, and then click on the Plot icon. Results are shown in Figure 4.12.

14. To create a report for this model, right-click on the Reports node in the Model Builder window and choose the level of the report (e.g., Brief Report). This will create Report 1 under the Reports. Rename Report 1 to Brief and click on it. The Report window will open, as shown in Figure 4.13. In this window, choose the desired report format (e.g., Microsoft Word) under Output format, choose where to save the report under Report directory, and size and type (e.g., small and JPEG) under Images. A Preview of the final report can be viewed before writing it by clicking the Write icon button.

48 • COMSOL for Engineers

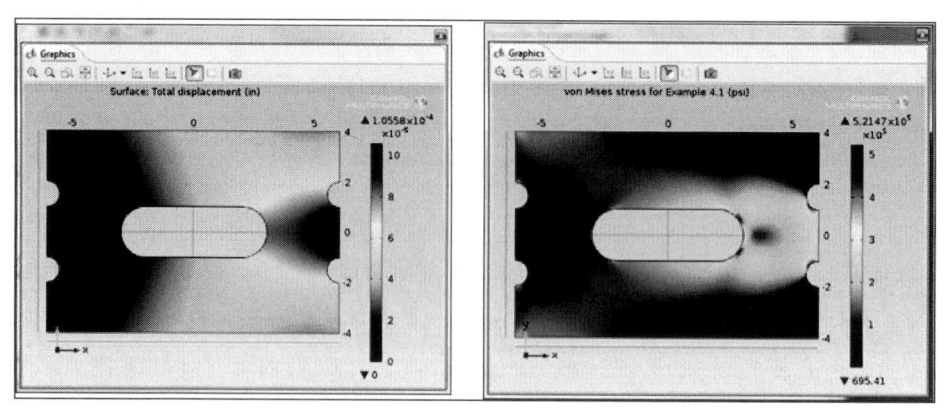

FIGURE 4.12 Results for displacement and von Mises stress.

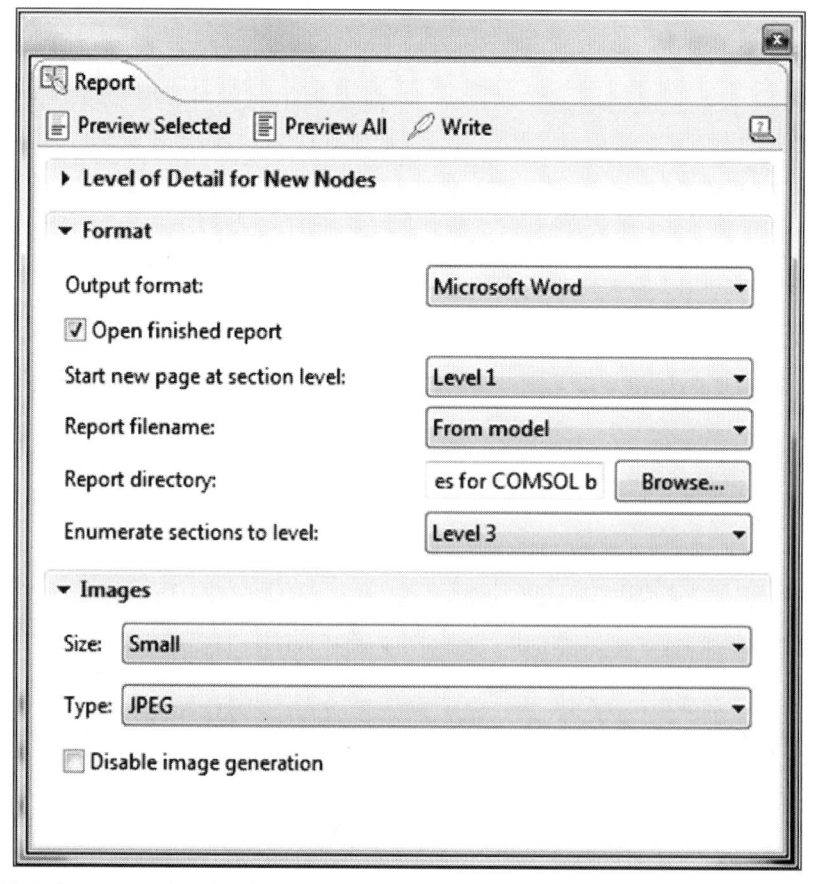

FIGURE 4.13 Data entry interface for creating a Report document.

Example 4.2: Dynamic Analysis for a Thin Plate: Eigenvalues and Modal Shapes

For the plate given in Example 4.1, we would like to calculate the eigenvalues (i.e., natural frequencies). Natural frequencies for a structure/machine are important characteristics in terms of its vibration analysis. For a given structure, in this example a plate, there exists a set of frequencies at which it can vibrate. The smallest value of this set is called the fundamental natural frequency of the structure. Natural frequencies are the solution of the homogeneous governing equations, and applied loads usually do not have an effect (or a very small effect on the global stiffness matrix) on the solution. However, boundary and initial conditions do affect the value of natural frequencies (see Reference 4.3).

In this example, we would like to calculate the natural frequencies of the plate given in Example 4.1 when initially at rest. Also we would like to calculate the corresponding modal shapes associated with natural frequencies.

Solution

We use the same model that we developed for Example 4.1, and add a Study scenario to it.

1. Open the model file Example 4.1. Save it as a new file with the name Example 4.2.

2. In the Model Builder window, open the list under Study 1 by clicking on the triangle icon on its left side. Right-click on the Study 1 node and choose Study Steps>Eigenfrequency. Step 2: Eigenfrequency will appear under Study 1 in the Model Builder window.

3. To run the model only for dynamic analysis, click on the Step 1: Stationary node under Study 1 in the Model Builder window. The Stationary window will open. Under the Physics and Variables Selection section, click on the check mark in the Solve for column in the table. An ✖ will appear, as shown in Figure 4.14.

4. For dynamic analysis, the density of the material is required. Open the list in Model 1 (by clicking on the triangle icon on its left) and click on Linear Elastic Material node under Plate node. In the Linear Elastic Material section, under Density open the list, choose User defined, and enter 7850. The unit used for density is kg/m^3, as shown in Figure 4.15.

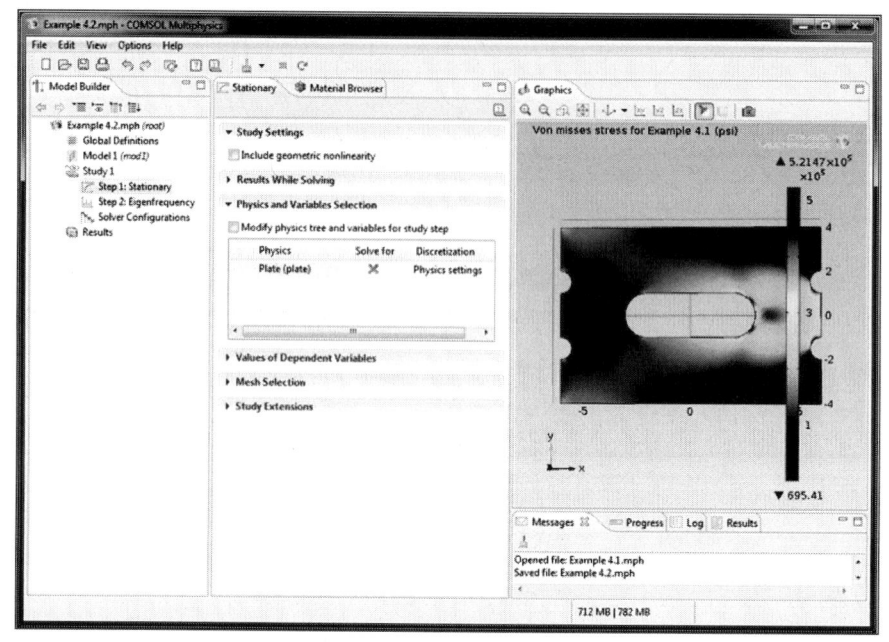

FIGURE 4.14 Results from Example 4.1 showing von Mises stress.

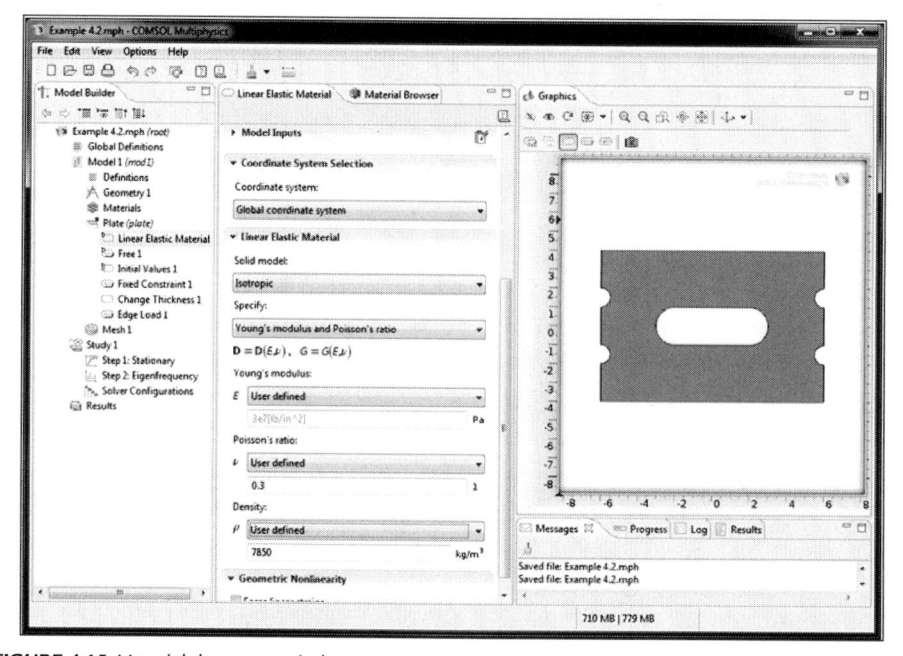

FIGURE 4.15 Material data entry window.

5. Click on the Study 1 node and run the model by clicking on the equal sign icon (≡) located in the toolbar. Wait until the run is finished.

6. Right-click on Stress Top (plate) node in the Model Builder, choose Rename, and rename it to 2D Plot Group. Click on the 2D Plot Group node to open the corresponding window. In this window, open the Eigenfrequency list in the Data section to see the calculated plate fundamental natural frequency (28.623534) and five higher natural frequencies, as shown in Figure 4.16.

7. Since we have added a new study to the model, we should modify the Graphics. Click on the Surface 1 node in the Model Builder to open the Surface window. In the Expression section under Unit, open the list and chose mm. Check the box for Description. Under the Title section, open the Title type list and choose Automatic. Click on Height Expression and in the corresponding window, under Unit choose mm. Click on the Plot icon to draw the graphics. By default COMSOL shows the fundamental natural frequency (the lower value of the frequencies of the plate) which is 28.624 Hz. The modal shape can be scaled relatively and arbitrarily. The result is shown in Figure 4.17.

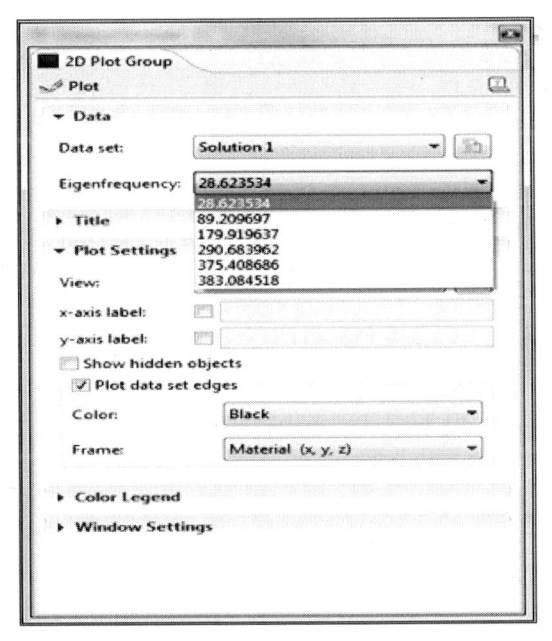

FIGURE 4.16 2D Plot data entry for Eigenfrequency.

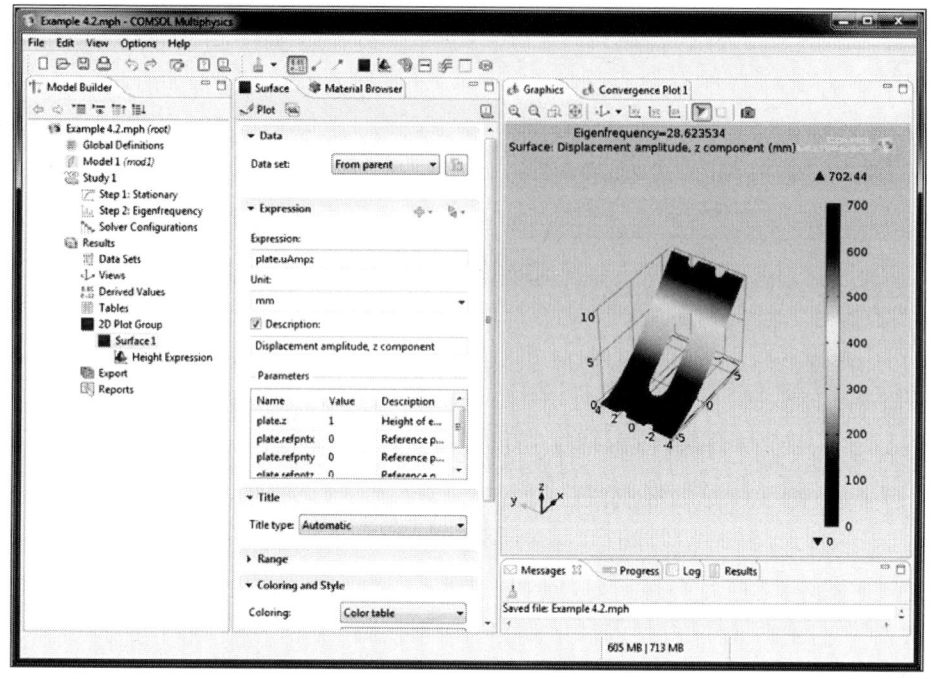

FIGURE 4.17 2D plot for eigenfrequency and displacement.

8. To see an animation of modal shapes, we can create a movie. Right-click on the Export node in the Model Builder window. A Player 1 node will be created and the Player window opens. In the Player 1 window, under Eigenfrequency list make sure all frequencies are highlighted/selected. In the same window, open the list in front of Frame selection and choose All. In the Playing section, check the Repeat box for the Display each frame option. Click the Generate Frame icon to create the Player 1. Result is shown in Figure 4.18.

9. To see the movie, click on the play icon located in the toolbar of the Graphics window. To stop the movie, click on the stop icon . It is recommended to stop the movie occasionally to see the mode shapes at different modes. Alternatively, a specific frame can be shown by moving the slider bar in the Frames section.

10. Use the Reports feature to generate a report for Example 4.2, following the procedure described in Example 4.1.

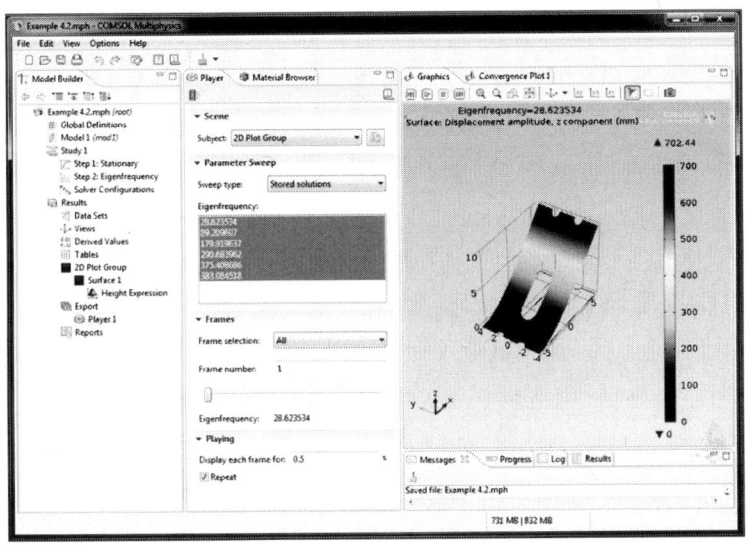

FIGURE 4.18 Data entry window for animation/Player for eigenfrequencies.

Example 4.3: Parametric Study for a Bracket Assembly: 3D Stress Analysis

In this example, we introduce Model Library, a useful resource in COMSOL. An example from this library (Structural_Mechanics_Module/Tutorial_Models/bracket_parametric)[1] will be rebuilt using COMSOL version 4.3. Parametric analysis, importing existing CAD files into a model, setting up global parameters also will be explained. The model also demonstrates 3D stress analysis available through COMSOL tools applications. This model has been treated extensively for different loading and analysis types in Model Library.

For analyzing stresses within a 3D solid body, we use the equilibrium equation (in tensor notation):

$$\frac{\partial \sigma_{ij}}{\partial x_j} + \chi_i = 0$$

---

[1] Model made using COMSOL Multiphysics® and is provided courtesy of COMSOL. COMSOL materials are provided "as is" without any representations or warranties of any kind including, but not limited to, any implied warranties of merchantability, fitness for a particular purpose, or noninfringement.

Where, σ_{ij} is the stress tensor that has nine elements in 3D space (i and j take values of *1, 2, 3*). For most cases in practice, we end up with a symmetric stress tensor and hence we would have six independent elements. The stress tensor can be related to the strain tensor and consequently to displacements using constitutive equations (like Hooke's law) and kinematic compatibility of strains constraint. The body loads per unit volume X_i, together with the divergence of stresses, should balance external loads and boundary conditions to have an equilibrium state for a given structure/machine. Applied loads may include static loads or dynamic ones. In the latter case, the vibrations of the structure can be analyzed, assuming a linear analysis, using modal shapes superposition. Each structure has a fundamental natural frequency under which it starts to vibrate. The fundamental natural frequency is a very important characteristic of a structure; if applied loads have the same frequency, then resonance will occur, displacements will become very large, and failure may occur as a result. For failure criteria, we use von Mises stress, the failure criterion for or limit for stresses based on maximum distortional energy. It assumes that yielding will occur when the distortional strain energy reaches that value which causes yielding in a simple tension test (see Reference 4.4).

For this example, we consider a bracket assembly made out of steel and held with bolts and which carries external loads applied through two pins, as shown in Figure 4.19.

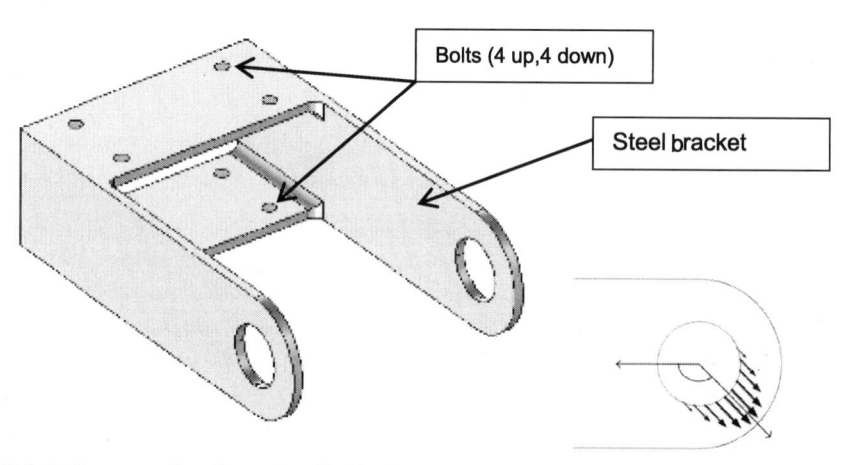

FIGURE 4.19 Geometry of bracket and applied-load pin holes.

External loads applied on the inner surface of the two holes in the bracket arms have a sinusoidal distribution $p = p_0 \sin(\alpha - \theta_0)$. We will find the resulting von Mises stress distribution.

Solution

1. Launch COMSOL and open a new file, and save it as Example 4.3.

2. In the Model Wizard window, choose 3D and then click on the arrow ⇨ to go to the next step. In the Add Physics window, open the list under Structural Mechanics, right-click on Solid Mechanics, and click on Add Selected to add it to the Selected physics list. Click on the Stationary node and click on the flag icon to Finish.

3. To set up global parameters, right-click on Global Definition in the Model Builder window and choose Parameters. The Parameters window will open and display a table. Enter the table values as shown in Figure 4.20. Information listed under Description column is optional.

4. To import the geometry file, right-click on the Geometry 1 node in the Model Builder window and select Import. The Import settings window will open. In this window, click Browse and browse to the folder

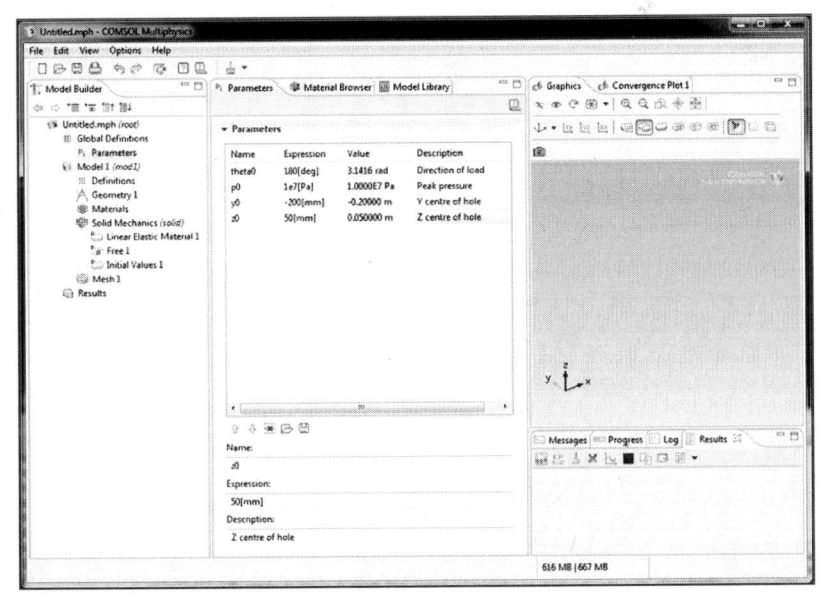

FIGURE 4.20 Parameters data entry.

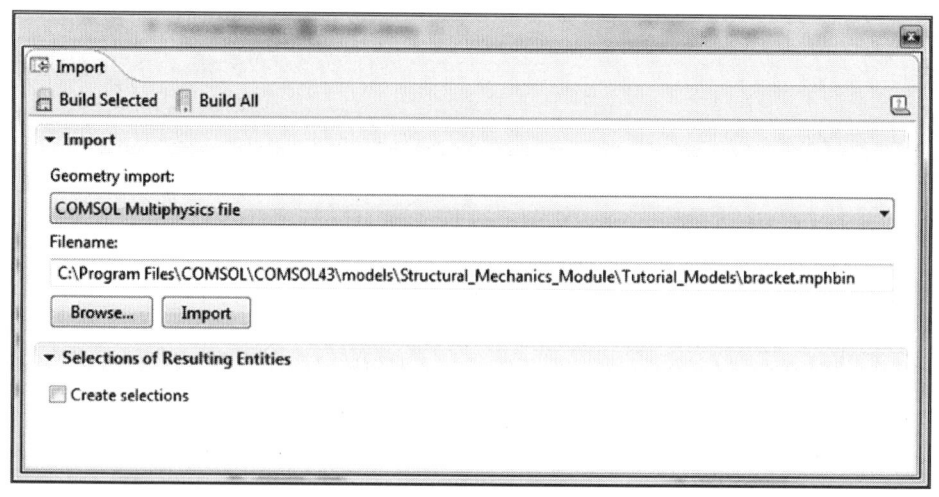

FIGURE 4.21 Importing a CAD file.

Structural_Mechanics_Module\Tutorial_Models. In the Tutorial_Models folder, double-click on the file bracket.mphbin, then click Import. (See Figure 4.21 for the complete file directory sequence.) You should have the bracket geometry as shown in the Graphics window.

5. Because this geometry is an assembly of the bracket and its bolts, we combine them to bond the bolts to the assembly. Otherwise the contact surfaces between the bracket and the bolts should be modeled by contact elements. In this example, we bond them together.

Right-click on the Form Union (*fin*) node in the Model Builder window and select Build Selected.

6. This model has complex loading on the bracket holes, specifically sinusoidal loads on half of the inner surface of the holes. To set up the loads we need to set variables and a local coordinate system. Right-click on the Definitions under Model 1 and select Variables. In the Variables settings window, enter the variables as shown as shown in Figure 4.22. (*Note*: Variables used should be exactly the same as those defined in the Parameters table.)

7. To set up the rotating coordinate systems for applied load on the holes, right-click on the Definitions node under Model 1 in the Model Builder window and select Coordinate Systems>Rotated System. The Rotated System window will open. In this window, enter (-theth0) for β, under Euler angles, as shown in Figure 4.23.

COMSOL MODELS FOR PHYSICAL SYSTEMS • 57

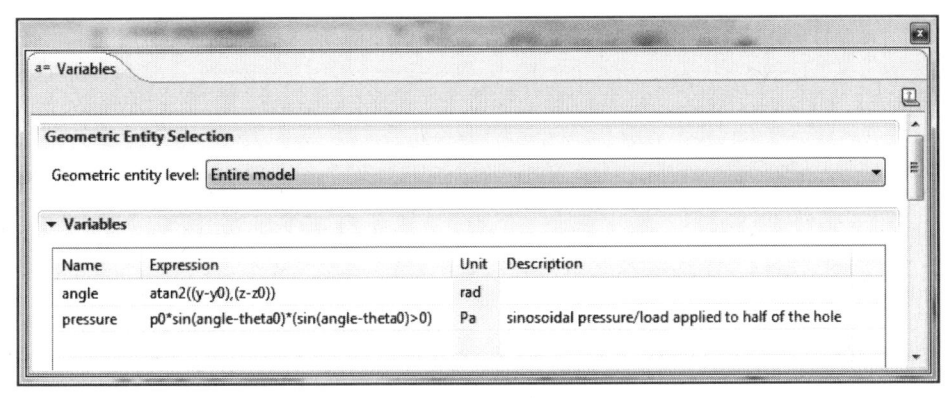

FIGURE 4.22 Data entry for Variables.

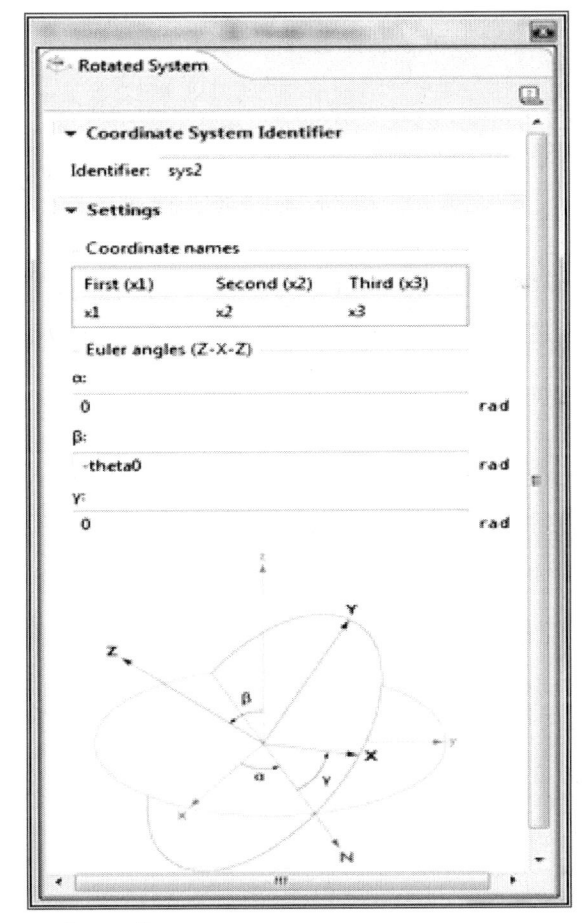

FIGURE 4.23 Data entry for coordinates.

8. To assign materials to the bracket geometry, right-click on the Materials node in the Model Builder window. The Materials window will open, as shown in Figure 4.24. In this window, open the list under Built-In, right-click on Structural Steel, and select Add Material to Model. Note that check marks will appear in the Material Contents section list for Density, Young's modulus, and Poisson's ratio (you may have to open this section by clicking on the pointer arrow). Although

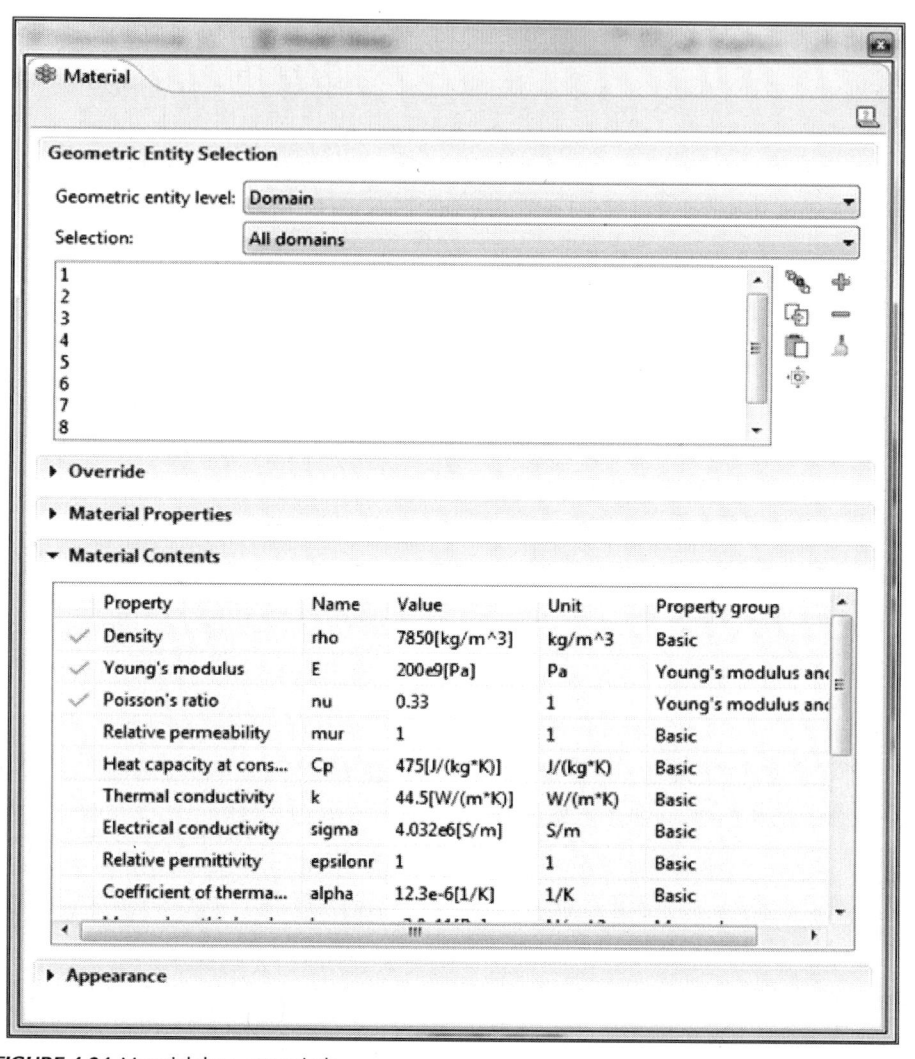

FIGURE 4.24 Material data entry window.

other material properties are listed, only these three are required for the type of physics involved in this model.

9. To assign the boundary conditions, right-click on the Solid Mechanics (*solid*) node in the Model Builder and select More>Fixed Constraint. The Fixed Constraint window will open. In the Graphics window, click on the bolts and add them to the list under the Selection by clicking on the plus icon ✚ , as shown in Figure 4.25. (You may have to rotate/zoom or click twice to choose the bolts only.).

10. To assign the applied load, right-click on the Solid Mechanics (*solid*) and select Boundary Load. The Boundary Load window will open. Click on the inner surfaces of the holes of the bracket in the Graphics window and add them to the Selection list. Under Coordinate System Selection, open the list to see the options and choose Rotated System 2. In the Force section, type **pressure** (should be exactly as defined in Variables 1) for the x_2 component. The symbols for applied forces will appear in the Graphics window. (To turn the physics symbols on, from the main

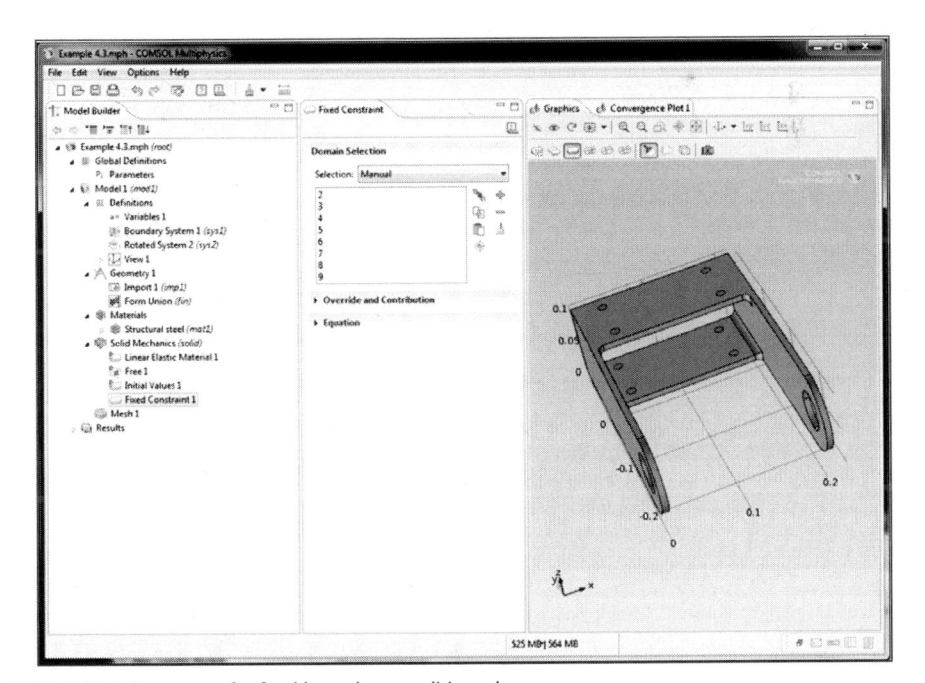

FIGURE 4.25 Data entry for fixed boundary conditions data.

menu select Options>Preferences and click the Graphics section. Click to select the Show physics symbols check box. Click Apply and OK. Click anywhere in the Model Builder, then click the node again. The symbols will be displayed in the geometry.) Results are shown in Figure 4.26.

11. Click on the Mesh node in the Model Builder. The Mesh window will open, as shown in Figure 4.27. In this window, select Normal for the Element size and click on the Build All icon [Build All]. Wait until the mesh is created and appears in the Graphics window.

12. To run the model, right-click on the Study 1 node in the Model Builder window and select Compute (≡). The Default result, von Mises stress distribution mapped on the deformed bracket, will appear in the Graphics window after the model run is finished, as shown in Figure 4.28. Maximum value is 39.5Mpa, which is less than yield stress value for structural steel (260 Mpa). This validates the choice of a linear elastic material and model to analyze this structure.

13. To show the direction of applied forces, right-click on the Stress (*solid*) node located under Results and select Arrow Surface. The

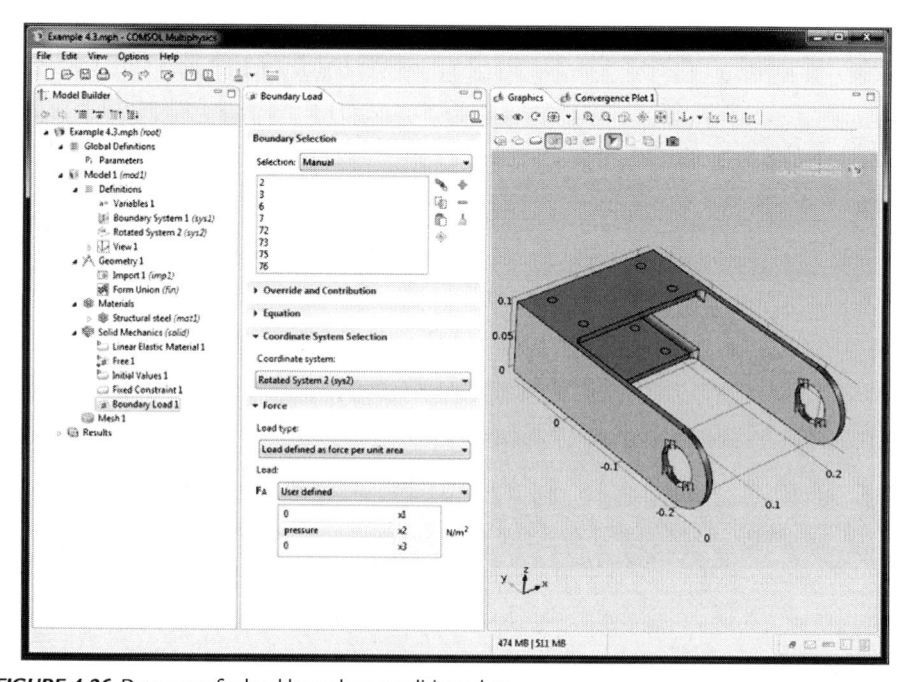

FIGURE 4.26 Data entry for load boundary conditions data.

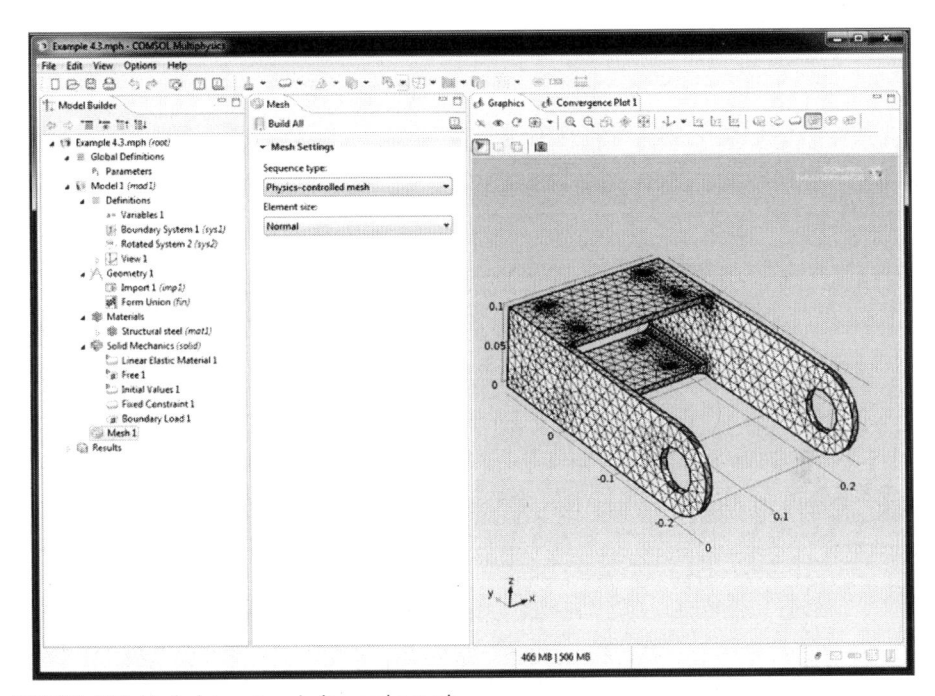

FIGURE 4.27 Mesh data entry window and a mesh.

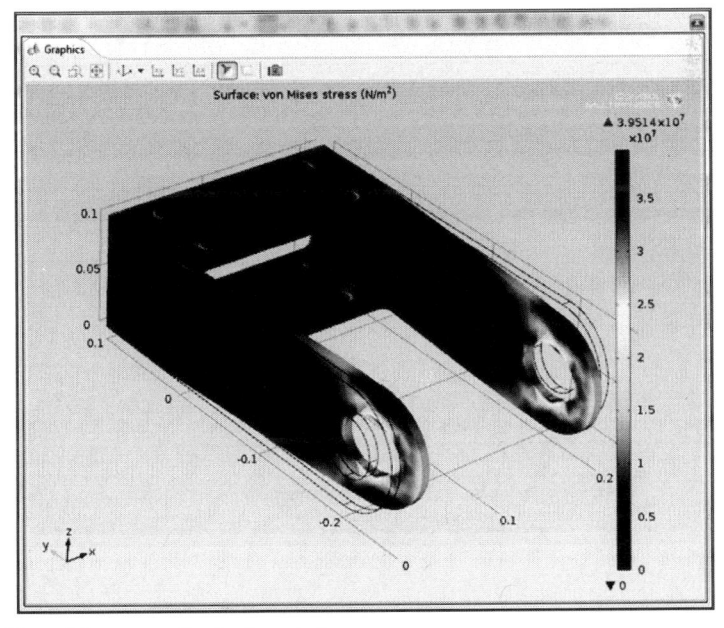

FIGURE 4.28 von Mises stress distribution window.

Arrow Surface window will open. In this window, click on the Replace Expression icon ( ) and select Solid Mechanics>Load>Load (Spatial)(solid.FperAreax, …, solid.FperAreaz) from the list. Under the Coloring and Style section, enter 3000 in the Number of arrows field. Click the Plot icon (Plot) to graph the results. Note that pressure arrows are shown on an undeformed Bracket. See Figure 4.29.

14. von Mises (σ_v) and principle stresses (σ_i) (are related as $\sigma_v = \sqrt{\dfrac{(\sigma_1-\sigma_2)^2 + (\sigma_2-\sigma_3)^2 + (\sigma_1-\sigma_3)^2}{2}}$. To show the principle stresses, right-click on Results and select 3D Plot Group. Right-click on the newly created 3D Plot Group 2 node in the Model Builder and select More Plots>Principal Stress Volume. The Principle stress volume window will open. In this Window, under Positioning section enter following data as shown in Figure 4.30 and then click the Plot icon (Plot).

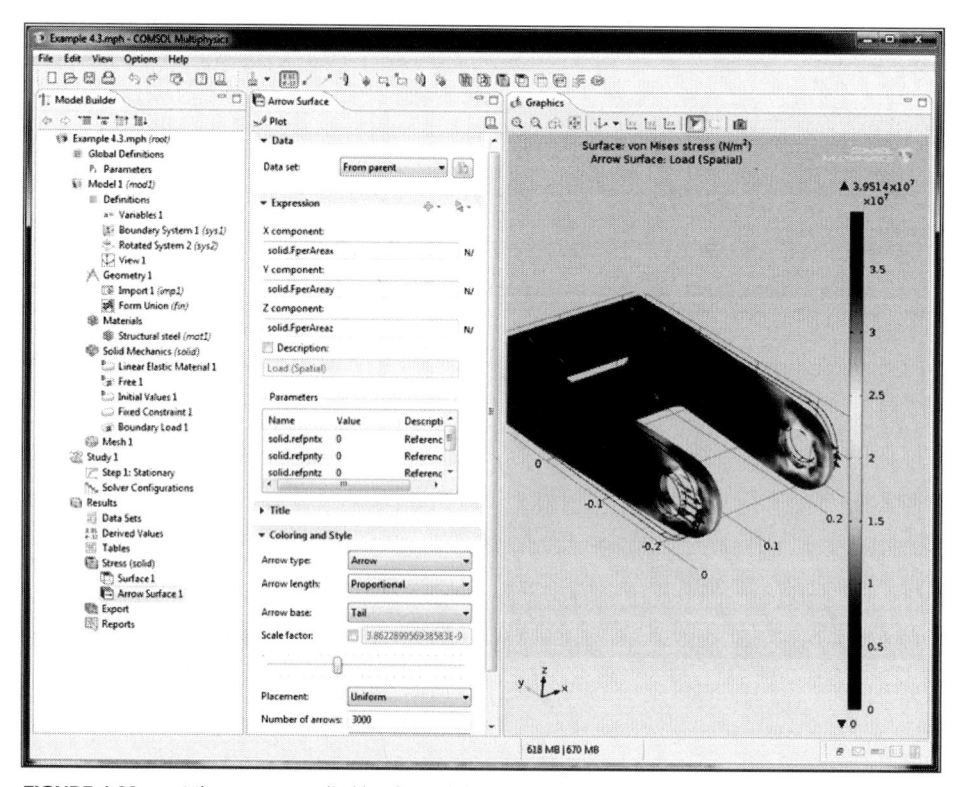

FIGURE 4.29 von Mises stress, applied loads, and deformed bracket geometry.

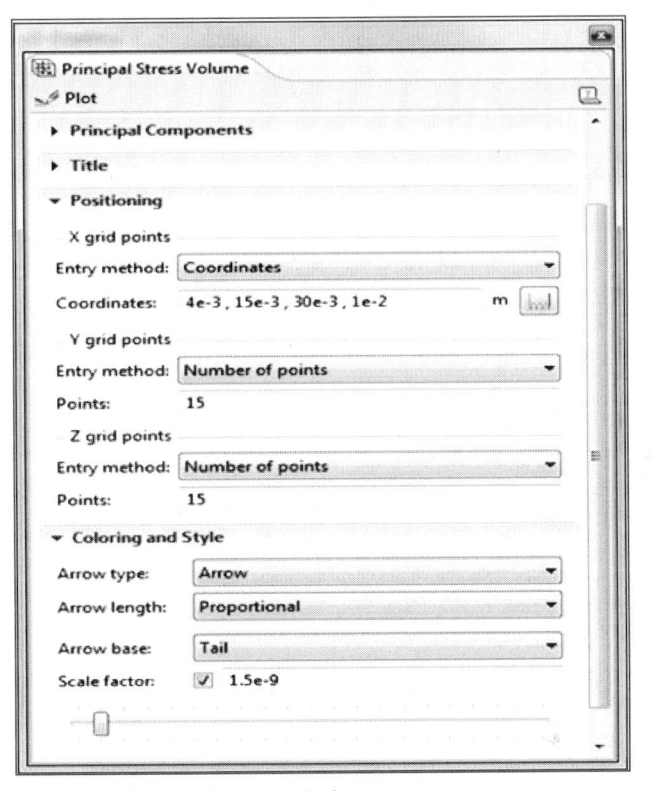

FIGURE 4.30 Principle Stress Volume data entry window.

15. In the Graphics window, click on the Zoom Box icon ( ) and zoom in on the left arm of the bracket. Principle stresses are shown with arrows (red the largest, green the medium, and blue the smallest—consistent with the coordinate axes).

16. It is useful to show the values of the reaction forces exerted by bolts. Right-click on the Derived Values node under Results in the Model Builder window and select Integration>Volume Integration. The Volume Integration window will open. In the Graphics window, click all bolts to select and add them to the list in the Selection section in the Volume Integration window. Under Expression, click on the Replace Expression icon ( ) and select Solid Mechanics>Reactions>Reaction force (Spatial)>Reaction force, x component (solid.RFx). Click the Evaluate icon ( ). In the Expression section, type solid.RFy and click Evaluate. Again, type solid.RFz in the Expression section and

click Evaluate. The Values of the components of reaction force vector appear in a table in the Results window (or select View>Results on the toolbar), as shown in Figure 4.31.

17. To study the effect of the orientation of the applied load, we perform a parametric study. The parameter is angle θ. Click Step 1: Stationary node (under Study 1 in the Model Builder window). The Stationary window will open. In this window, click on the Study Extension section to expand it. Check the Continuation box. Under Continuation parameter, click the ✚ icon to add the parameters to the table. Open the list (which shows up under Continuation parameter) and select thetah0 (Direction of load). Under Parameter value list, type **range(0,45[deg],180[deg])**. This will change the value of parameter theta0 from 0 to 180 degrees with increments of 45 degrees. See Figure 4.32.

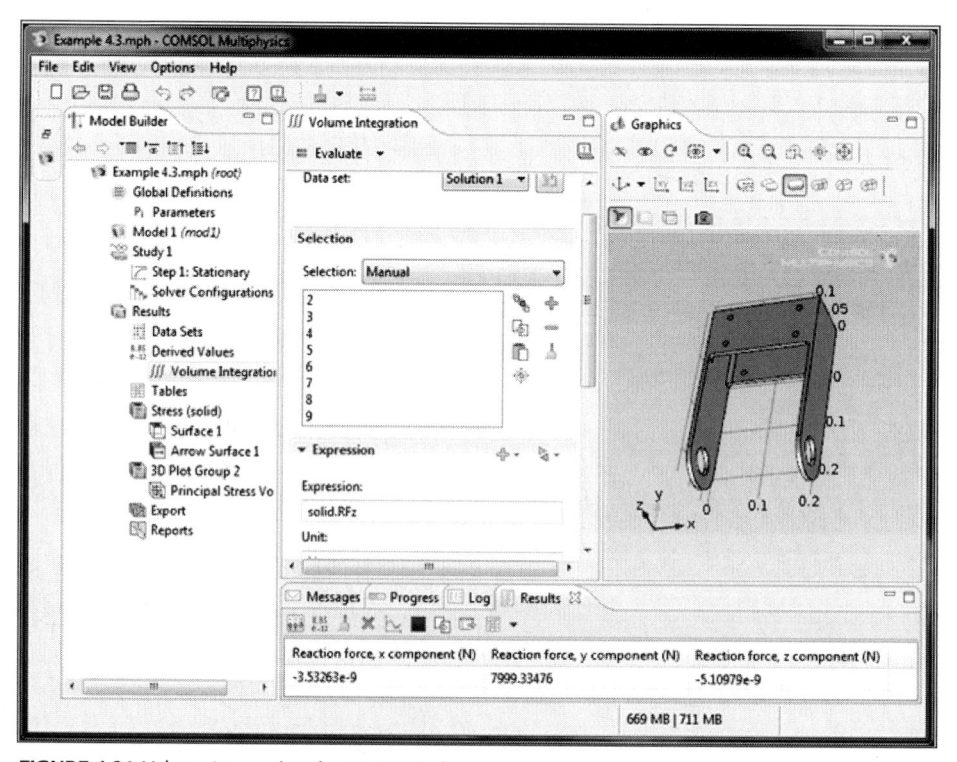

FIGURE 4.31 Volume Integration data entry window.

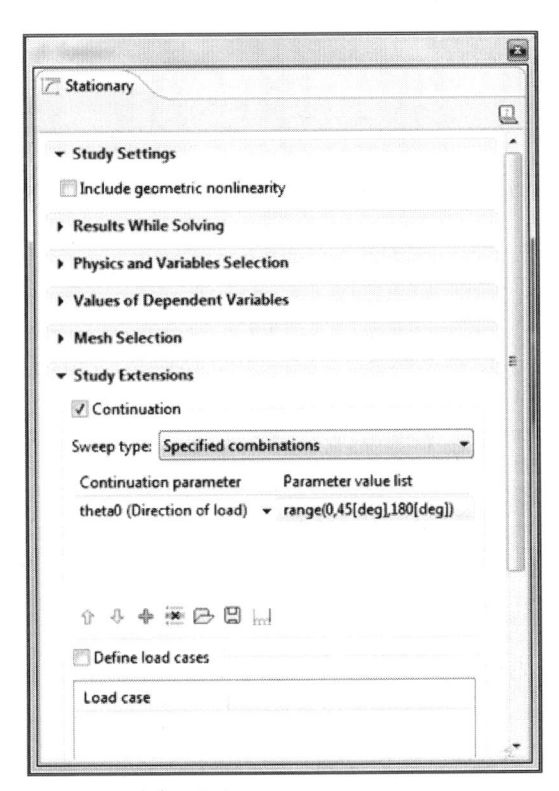

FIGURE 4.32 Study Extension data entry window.

18. Right-click the Study 1 node (in the Model Builder window) and select Compute ≡ . Wait for the model run to finish the calculations. The results will appear in the Graphics window. Click Stress (*solid*) node in the Model Builder and, from the corresponding window, under Data section, select 0 from the list of Parameter values (**theta0**) and click the Plot icon to plot it. Note that the direction of applied forces (shown by red arrows) changes. Similarly, choose another value for the list (e.g., 1.570796) for **theta0** and plot it. The result is shown in Figure 4.33.

19. To calculate the reaction for different orientations of the applied load, right-click the Derived Values node (under Results in the Model Builder window) and select Integration>Volume Integration. Repeat the operations similar to those explained in Step 16 above. Results for reaction forces will appear in a table, as shown in Figure 4.34.

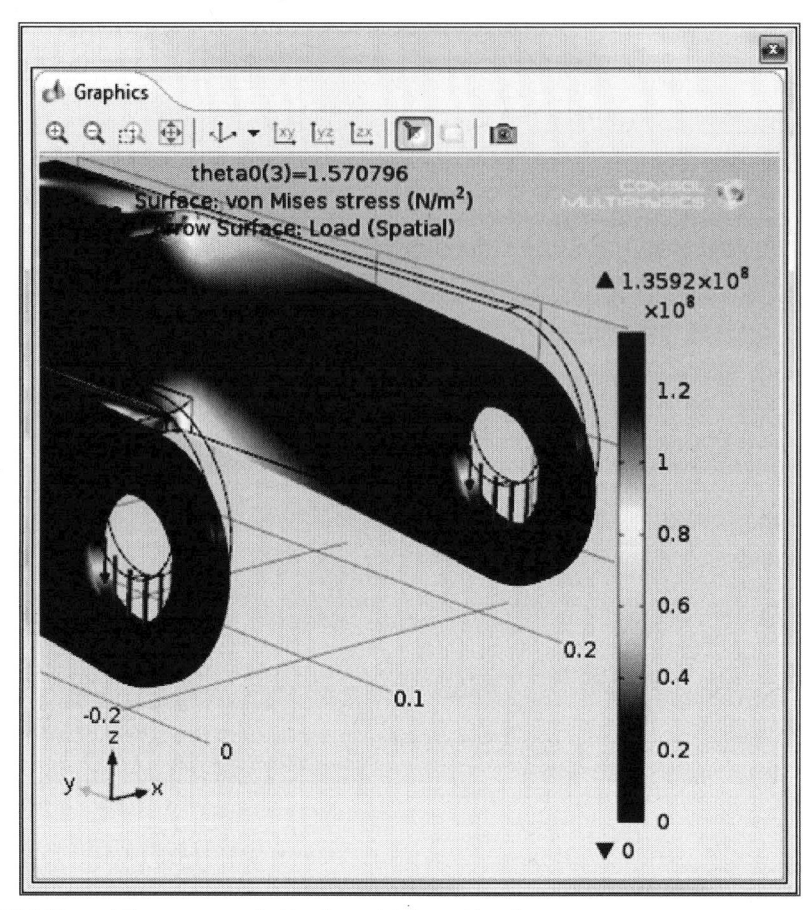

FIGURE 4.33 von Mises stress, applied loads, and deformed bracket geometry.

theta0	Reaction force, x component (N)	Reaction force, y component (N)	Reaction force, z component (N)
0	2.4797e-9	-7999.32877	6.06417e-9
0.7854	-2.45922e-8	-5646.24486	5646.24486
1.5708	-3.97271e-8	2.33527e-8	7999.33177
2.35619	-3.1395e-8	5646.24785	5646.24785
3.14159	-2.49337e-9	7999.33476	-4.88234e-9

FIGURE 4.34 Results for derived reaction forces.

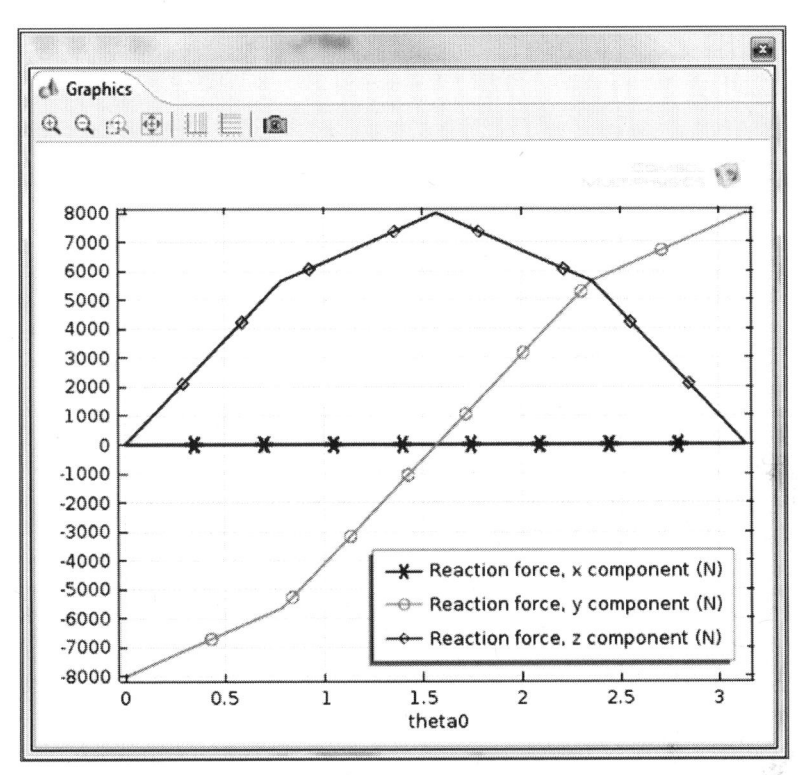

FIGURE 4.35 Reaction forces for different theta0, orientation angle.

20. It would be useful to make a graph of reaction forces. Right-click the Results node (in the Model Builder window) and select 1D Plot Group. In the 1D Plot Group new window that opens, expand the Legend section and choose Lower right from the list. Right-click on the 1D Plot Group 3 and select Table Graph. A new window will open. In this window under Data, locate Table, open the list, and select Table 2. Under the Legends section, check the Show legends box (you may need to expand this section to view the options). Click Plot. A graph showing the values of reaction forces (in N) versus theta0 (in rad) will appear in the Graphics window. Results are shown in Figure 4.35.

Example 4.4: Buckling of a Column with Triangular Cross-section: Linearized Buckling Analysis

A column is a structural member that supports applied load, mainly in compression. For certain values of an applied load and boundary conditions, the column may fail and exhibits very large deformations. The failure situations are examples of instability from neutral equilibrium condition for the given column geometry, material, and boundary conditions.

Buckling is categorized as a bifurcation problem from a mathematical point of view since it has more than one equilibrium situation when and after the column becomes unstable. One important point to emphasize here is that buckling is not a material failure type; it is the result of the column becoming unstable under a given load. A column could buckle for discrete values of applied loads. The smallest value—clearly of interest to engineers—is called the critical load (Euler formula: for a simply supported column with length L and moment of inertia I, $P_{cr} = \pi^2 EI / L^2$). Buckling could also be considered an eigenvalue/eigenvector problem, and is one of the approaches available in COMSOL. Readers interested in more in-depth discussions on buckling are referred to Reference 4.4 and the COMSOL help manual.

In this example, we use a linearized buckling analysis that provides an estimate of the critical load that causes sudden collapse of the column. We calculate the critical load for a column with an equilateral triangular cross-section (side length is 30 cm) and a circular hole with radius 4 cm with its center located at x = 15 cm, and y = 12cm. For material properties, we use the existing material library in COMSOL, Aluminum 6063-T83 with Density 2700 kg/m^3, Young's modulus 69E9 Pa, and Poisson's ratio 0.33.

Solution

1. Launch COMSOL and open a new file.

2. Select 3D and click ⇨ Next. In the Add Physics window, open the list for Structural Mechanics, right-click on Solid Mechanics (*solid*), and click Add Selected to add it to the Selected physics list. Then click ⇨. In the Select Study Type window (under Preset Studies list) select Linear Buckling and click on the flag icon ⚑ to Finish.

3. Right-click on Global Definitions in the Model Builder window and select Parameters. Enter **col_height** under Name in the Parameters window/table and 200[cm] under Expression.

4. Geometry of the column can be built in COMSOL or alternatively imported as a file. In this example we build it using 3D CAD tools available in COMSOL.

4.1 Right-click on the Geometry 1 node in the Model Builder window and select Work Plane from the list. The Work Plane window will open. Click on the Plane Geometry node (under Work Plane 1). Click on the Draw Line ( ) icon in the toolbar and draw a horizontal line in the Graphics window. (Click at a point in the drawing area, move the mouse pointer to a second point, and click. Right-click to finish). Under Plane Geometry, click on Bezier Polygon 1 node (you may have to open the list). The Bezier Polygon window will open. In this window, click on Segment 1(*linear*) in the Polygon Segments section. Change the coordinates of the line to (0,0) and (30[cm],0), as shown in Figure 4.36, and click Build Selected ( ), as shown in Figure 4.37.

4.2 In the same window, click on Add Linear and enter the coordinates (30[cm],0) and (15[cm], 0.15*3^0.5). Click the Build Selected icon . Periodically click on the Zoom Extents ( ) icon to see all the geometry in the Graphics window.

FIGURE 4.36 Line point coordinates data entry.

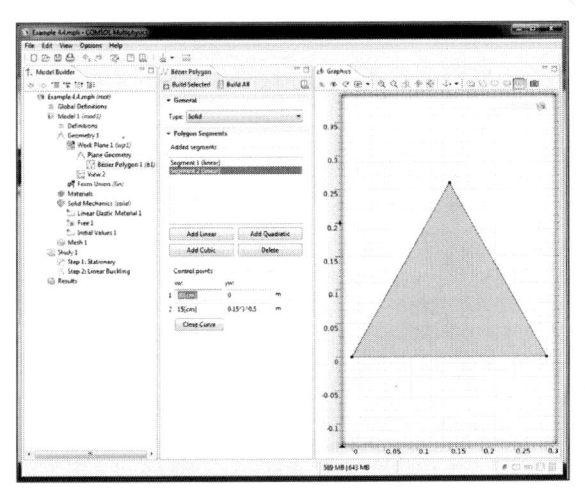

FIGURE 4.37 Control points data entry and resulting polygon geometry.

4.3 To add fillets, in the Model Builder window right-click on the Plane Geometry node and select Fillet. In the Fillet window, add the three vertices of triangle to the Vertices to fillet list by clicking on each one and then right-clicking (or clicking on ✚). Enter 1[cm] for the Radius and click 🔲 Build Selected. The filleted triangle will appear in the Graphics window, as shown in Figure 4.38.

4.4 To create the circle, draw a circle inside the triangle by right-clicking on the Plane Geometry node and selecting Circle. In the Circle window, enter **4[cm]** for the Radius, **15[cm]** for xw, and **12[cm]** for yw. Click Build Selected 🔲. To create the hole inside the triangle, click on the Select Object icon 🔲 (located on the Graphics window toolbar) and Alt+Click at any point in the Graphics area. The entire geometry is selected and the color turns red. Click on the Plane Geometry node again and click on the Difference icon 🔲 located in the main toolbar. The triangle with a circular hole appears in the Graphics window, as shown in Figure 4.39.

4.5 To extrude the cross-section, right-click on Work Plane 1 and select Extrude. The Extrude window opens. In this window under Distance, enter **col_length** and click on Build Selected icon. Click on Zoom Extents (🔲) in the Graphics window to see the entire Column geometry. See Figure 4.40.

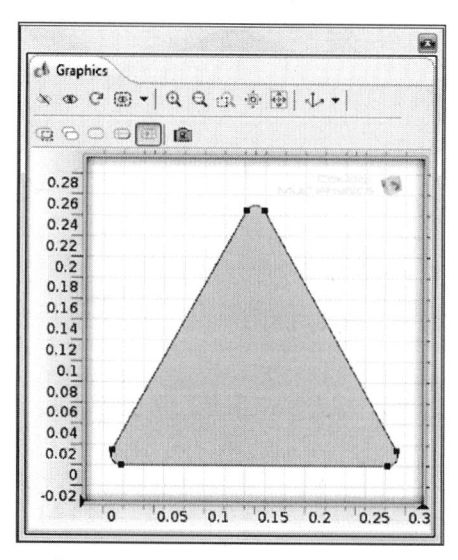

FIGURE 4.38 Column cross-section geometry.

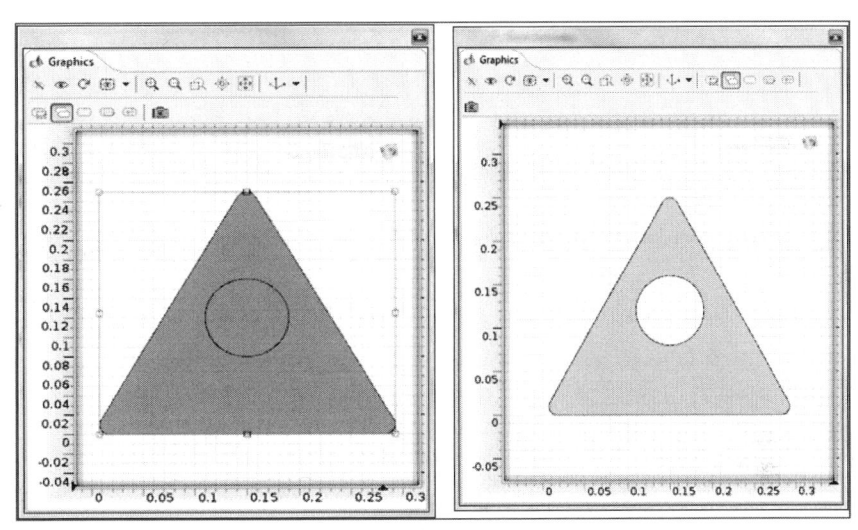

FIGURE 4.39 Column cross-section geometry with subtracted circular hole.

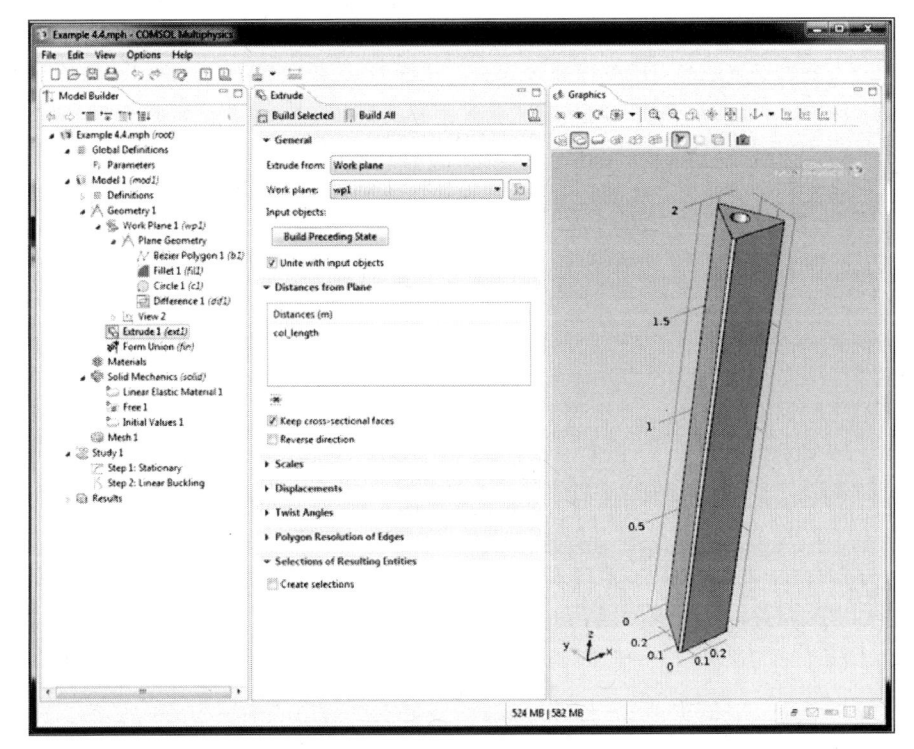

FIGURE 4.40 Graphics window showing column geometry.

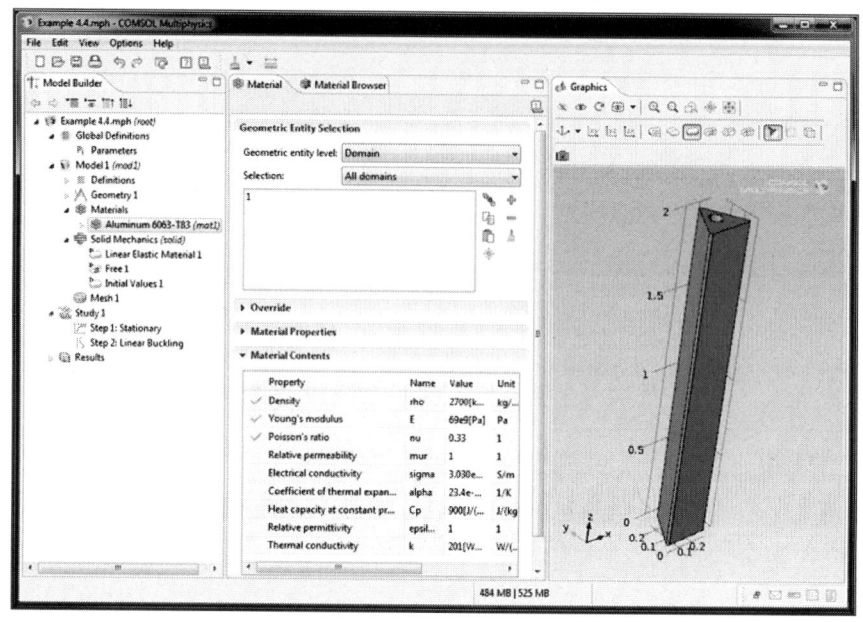

FIGURE 4.41 Material data entry for column.

5. To assign materials to the geometry, right-click on the Materials node in the Model Builder and select Open Material Browser. In the Material Browser window, open the list for Built-In, right-click Aluminum 6063-T83, and select Add Material to Model. The Material window will open. To assign the selected material to the column, click on the geometry (any point) and then right-click (or click ✚ in the Material window). The domain number should be listed under the Selection in the Material window. See Figure 4.41.

6. Define a global parameter for the applied load on the column. Click on the Parameters node in the Model Builder window. In the Parameters window table, type **col_load** under Name and **100[Pa]** under Expression, as shown in Figure 4.42.

7. To define the boundary conditions , right-click on the Solid Mechanics (*solid*) node in the Model Builder window and select Fixed Constraint from the list. Assign this constraint to the base of column by clicking on the base surface (located in the x-y plane) of the column, and add it to the Selection list by right-clicking again (or clicking on the ✚ icon). Users may need to rotate the column geometry to see the base.

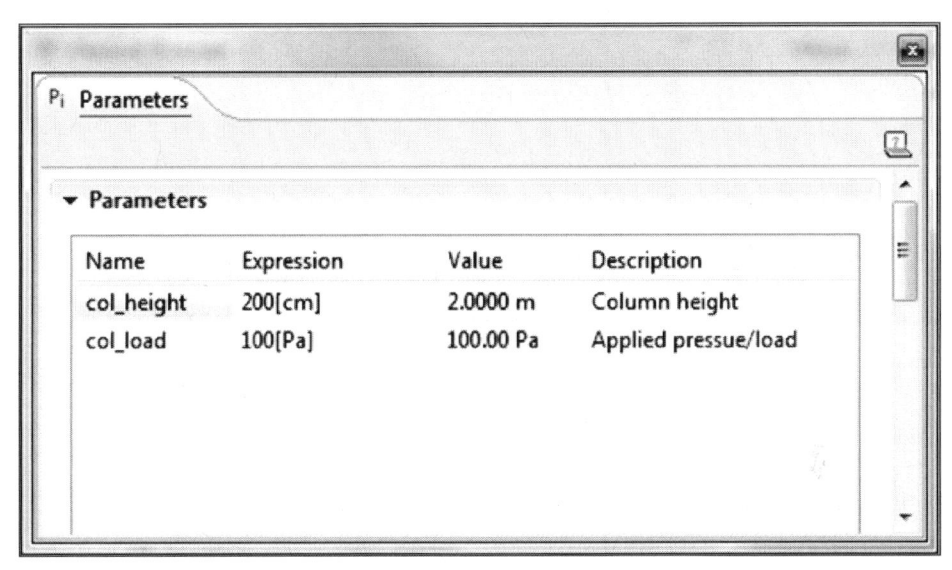

Name	Expression	Value	Description
col_height	200[cm]	2.0000 m	Column height
col_load	100[Pa]	100.00 Pa	Applied pressue/load

FIGURE 4.42 Parameters data entry window.

To define the applied load on the top surface of the column, right-click on the Solid Mechanics (*solid*) node and select Boundary Load from the list. Assign this load to the top of the column by clicking on the top surface (located in the z = col_height plane) of the column, and add it to the Selection list by right-clicking (or clicking on the ✚ icon). In the Boundary Load window, locate Force section and enter **–col_load** (the minus sign is needed since load is applied in the negative z-direction) for the z component, as shown in Figure 4.43.

8. To run the model, right-click on the Study 1 node and select Compute (≡). COMSOL will build a mesh automatically using default element sizes. The solver will first solve the equilibrium equations and then find the first eigenvalue or the critical load factor. The result for first mode shape of buckling and value of critical load factor (1.685879E6) is shown in the Graphics window.

9. To calculate the critical load of buckling for this column with assigned load factor and boundary condition, use $P_{cr} = \lambda \times col\_load$, where λ is the critical load factor.

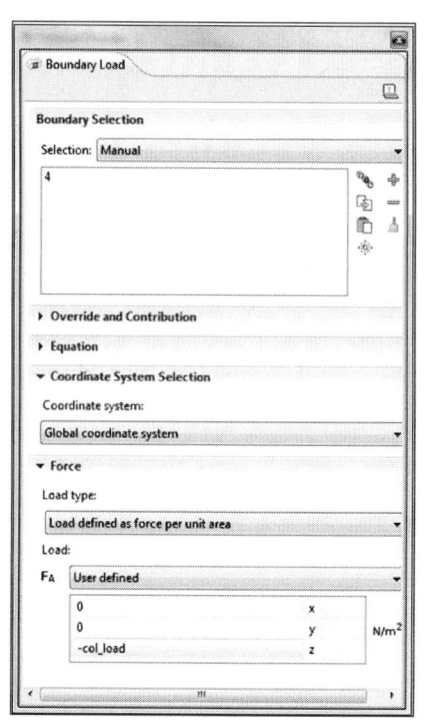

FIGURE 4.43 Boundary Load data entry.

10. To perform a parametric analysis for the applied loads, right-click on the Study 1 node in the Model Builder and select Parametric Sweep. The Parametric Sweep window opens. In this window under Parameter names, click on the ➕ icon to display a list and select col_load (Applied pressure/load). In the space under Parameter value list, enter **100, 200, 300**, as shown in Figure 4.44. These are the values for applied load on the column. To run the model, right-click on the Study1 node and select Compute.

11. The Results appear in the Graphics window. Click on the Mode Shape (*solid*) 1 node in the Model Builder (under Results). In the corresponding window, select Solution 3 from the list in front of Data set. Open the Parameter value (col_load) list (the three values of applied load are listed here). Choose, for example, 200. The corresponding critical load factor will appear in the Critical load factor (space under the Parameter value). To plot this result in the Graphics window, click on the Plot ✎, as shown in Figure 4.45. Similarly, the results for other loads can be obtained and plotted.

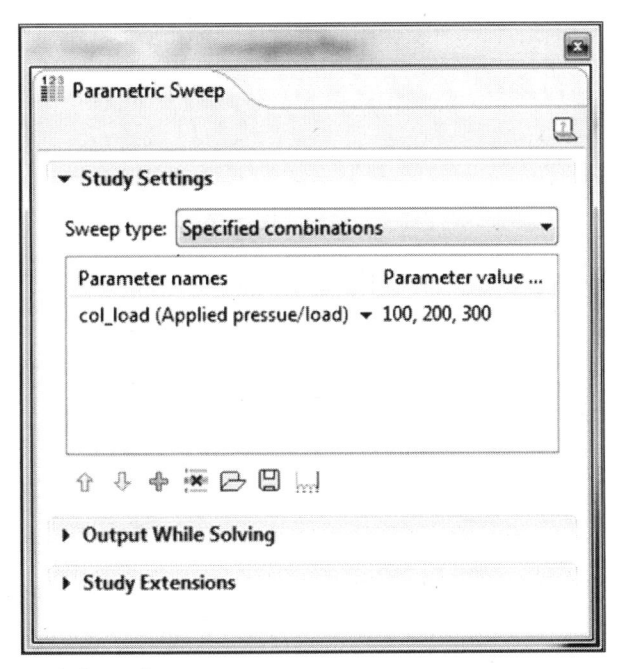

FIGURE 4.44 Parametric Sweep data entry.

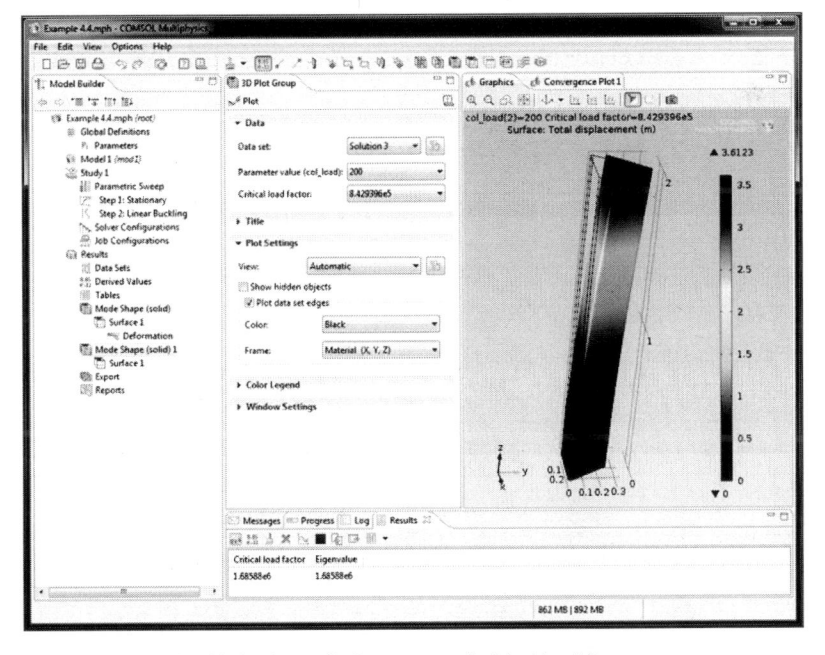

FIGURE 4.45 Results for buckled column displacement and critical load factor.

Example 4.5: Static and Dynamic Analysis for a 2D Bridge-support Truss

A truss is a structure that carries loads applied to it and results in compression or tension in its members. In this example, we use tools available in COMSOL to analyze a typical bridge structure under static and dynamic loads, including calculating its fundamental natural frequency. Truss dimensions (in feet) are given in the Figure 4.46.

Solution

1. Launch COMSOL and open a new file. Save it as **Example 4.5**.
2. In the Model Wizard window, select 2D button and click on Next ⇨.
3. From the Add Physics list, expand Structural Mechanics node and from the list select Truss (*truss*) and add it to the Selected physics list by clicking on ✚. Then click Next ⇨, as shown in Figure 4.47.
4. From the Select Study Type, select Stationary and click Finish ⚑. At this stage the physics, type of study, and level of dimension (2D) of the model are set for the truss.

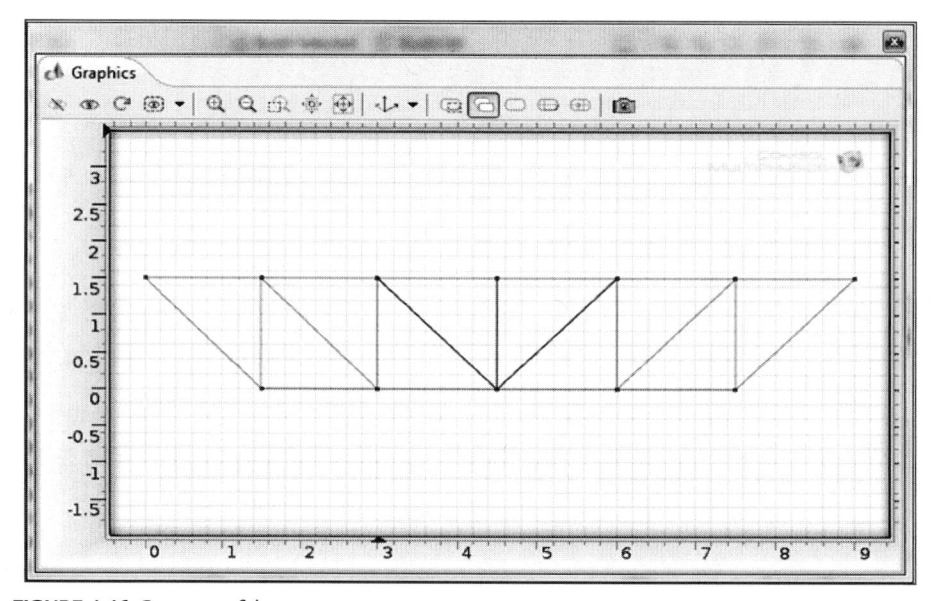

FIGURE 4.46 Geometry of the truss.

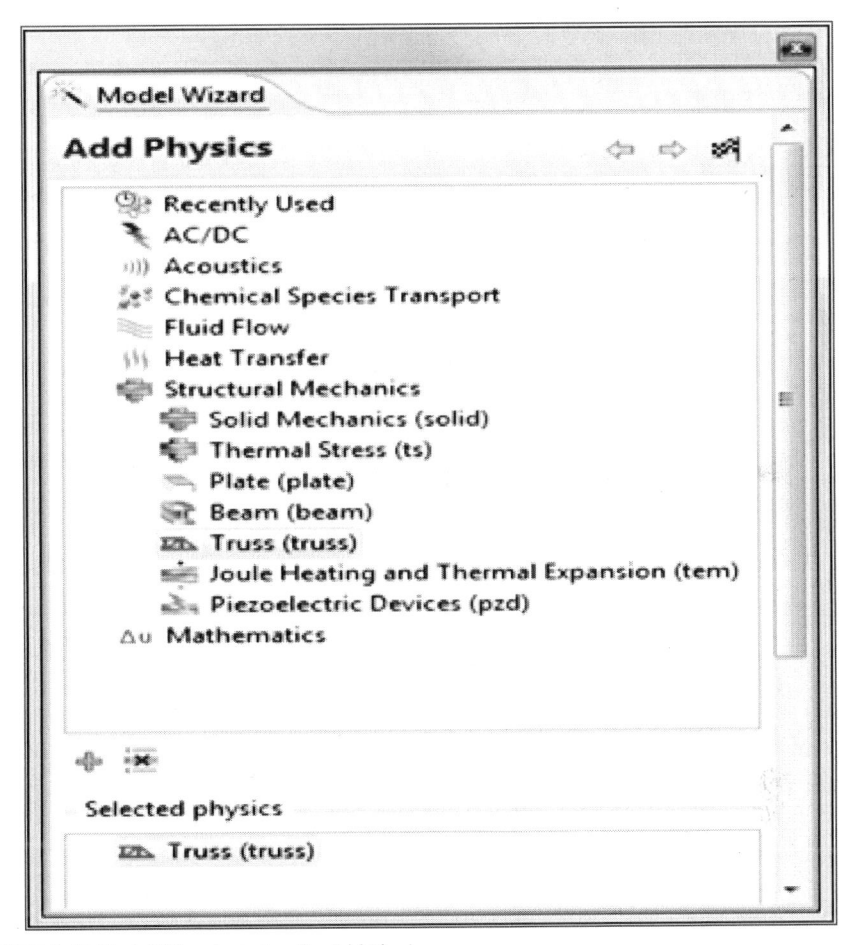

FIGURE 4.47 Model Wizard window for Add Physics.

Now we draw the geometry of the truss in the Graphics window using CAD tools available in COMSOL.

5. Since the dimensions are given in feet, change the Length unit to ft, from the list in Units section in Geometry window. Note that all units will be automatically converted to the SI units. Click on the Draw Line icon located in the main toolbar and draw a horizontal line in the Graphics window. To draw a line, click on a point then drag the mouse to another point and then click again (to release the draw line tool, right-click). In the Model Builder window, expand the Geometry node

and click on the Bezier Polygon 1(*b1*). In the corresponding window, click on Segment 1 (*linear*) and enter coordinates for x: 0 and y: 1.5, x: 9 and y: 1.5. Click Build All.

6. Using the same tool, draw the rest of truss members, as shown in Figure 4.48, using the following data:

Node	A	B	C	D	E	F	G	H	I	J	K	L
x: (ft)	0	1.5	3	4.5	6	7.5	9	1.5	3	4.5	6	7.5
y: (ft)	1.5	1.5	1.5	1.5	1.5	1.5	1.5	0	0	0	0	0

7. To assign material to the model, right-click on the Materials node in the Model Builder and select Open Material Browser. In the corresponding window, as shown in Figure 4.49, expand Built-In and select Structural steel from the list. Click on ✚ to add material to the model. Numerical values of material properties will be listed in the Material Contents section. These values can be modified, if needed. In this example, we accept the default values.

We have built the truss geometry and assigned material properties. We now define boundary conditions, applied loads, and truss members shape properties. We refer to the truss sketch for nodes.

8. Right-click on the Truss (*truss*) node in the Model Builder window and select Cross Section Data from the list. In the corresponding window, locate Basic Section Properties and enter **3e-3** for Area: A. This is equal to the area of two 4 x 4 x 5/8" in. angle shape steel members. Create another Cross Section Data node and enter **2*3e-3** for its Area value.

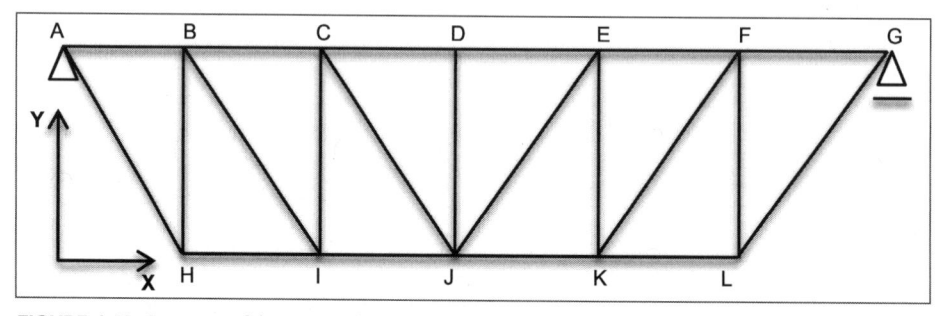

FIGURE 4.48 Geometry of the truss and its joints coordinates.

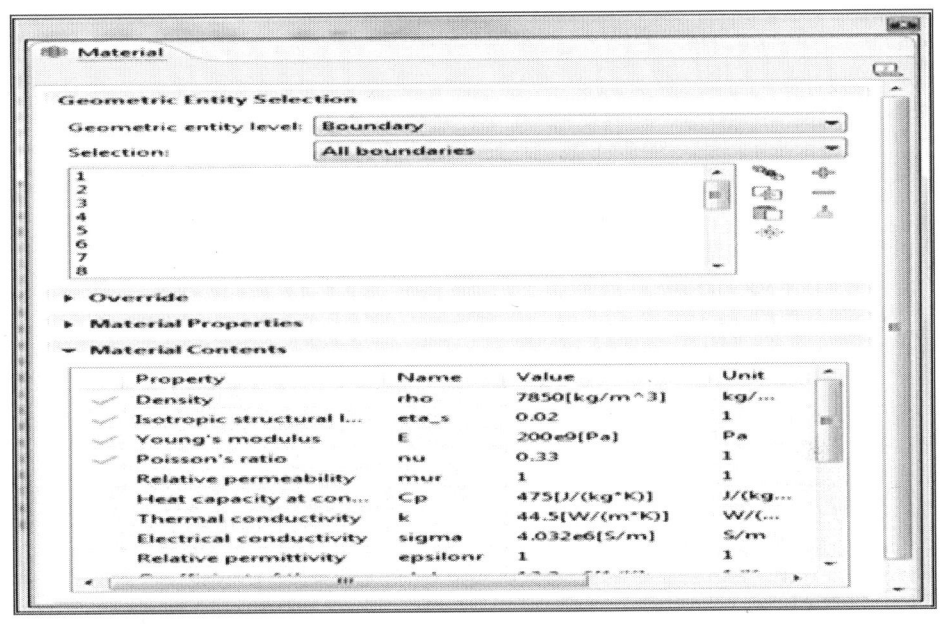

FIGURE 4.49 Material properties for truss members.

9. To add point loads, right-click on the Truss (*truss*) node and select Point Load from the list. In the corresponding window, as shown in Figure 4.50, select and add the top middle node (node D) to the Selection list. Locate section Force and enter 0 for x and −8e3 for y. Similarly, create another Point Load and select nodes B and C to it. For the load values, enter 2e3*0.1 for x and −2e3 for y.

10. For boundary conditions, right-click on the Truss (*truss*) node again and select Pinned from the list. In the corresponding window, as shown in Figure 4.51, add node A to the Selection list. This will assign a pinned support condition for this node which has x = y = 0 for its displacements. For the other support node G, we constrain displacement in y-direction to be zero, y = 0. Right-click on the Truss (*truss*) node and select Prescribed Displacement from the list. From the Graphics window, select the corresponding node for node G and add it to the Selection list in the Prescribed Displacement window under Point Selection section. In the same window, locate the Prescribed Displacement section and check the box for Prescribed in y direction, only. Make sure the value for V_0 is set to 0.

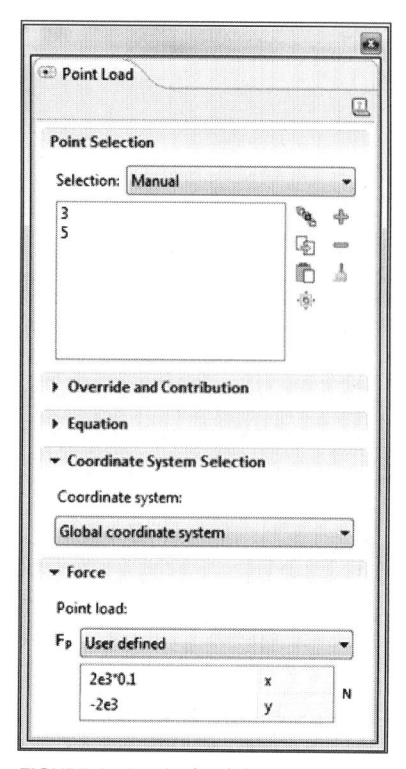

FIGURE 4.50 Point load data entries.

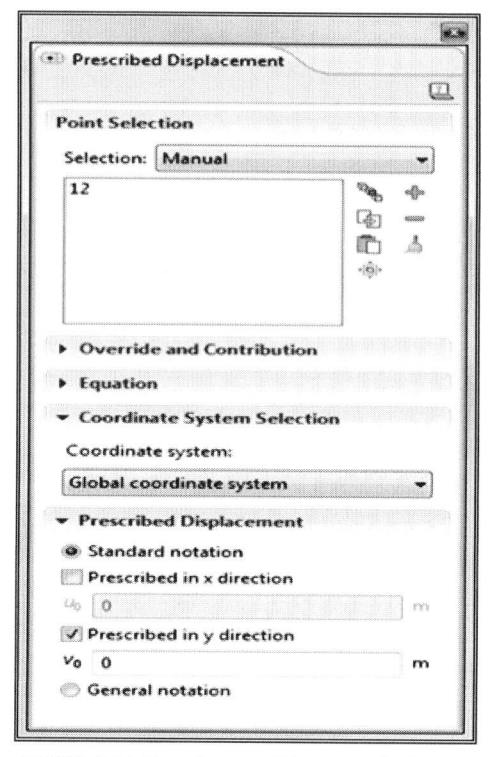

FIGURE 4.51 Boundary condition prescribed displacement data entries.

At this stage, we have the model ready for meshing. For truss structures a low-resolution mesh would be sufficient since each member can actually be considered as an element.

11. Click on the Mesh 1 node in the Model Builder window and in the corresponding window select Extra coarse from the list for Element size. Click Build All .

12. To run the mode, right-click on the Study 1 node in the Model Build window and select Compute . Wait for the computations to finish.

13. Default results for normal forces for members will appear in the Graphics window, as shown in Figure 4.52, corresponding to Force (*truss*) node under the Results. Also, normal stresses for truss members can be shown by clicking on corresponding node Stress (*truss*).

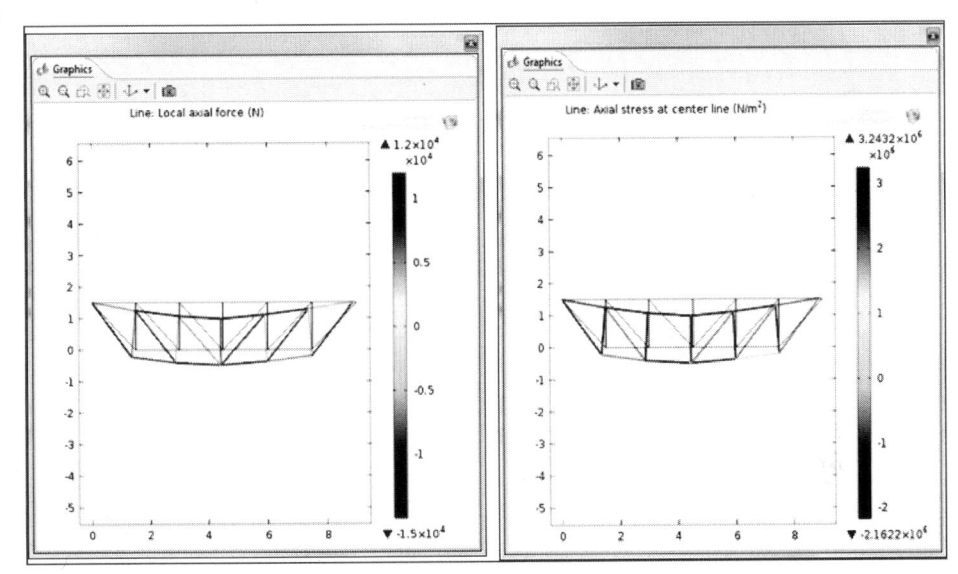

FIGURE 4.52 Results for truss members axial loads and stresses.

14. To show the displacement results, right-click on Results and select 2D Plot Group. A new 2D Plot Group node will appear under the Results tree. Rename to the new node to Displacements. Right-click on Displacements and select Line. In the corresponding window, click on Replace Expression icon and select Truss>Displacement>Total displacement (*truss.disp*). For the Unit select **mm** from the list. Click Plot. To show the deformed shape, right-click on Line1 and select Deformation. Results are shown in Figure 4.53.

At this point static analysis is complete. For design purposes, a user can modify the loads or materials and design the truss to meet a design Code criterion.

Now we perform dynamic analysis by adding studies to the existing model. To start, we calculate the natural frequencies of the truss. Usually the natural frequencies are calculated without any applied load, hence the results are solution to homogeneous form of the governing equations. For some structures, compression or tension forces may affect the natural frequencies. For comparison, we calculate the natural frequencies with applied loads as well.

15. First rename the Static study case and its corresponding Solver 1 to Study1-Static and Solver 1-Static, respectively.

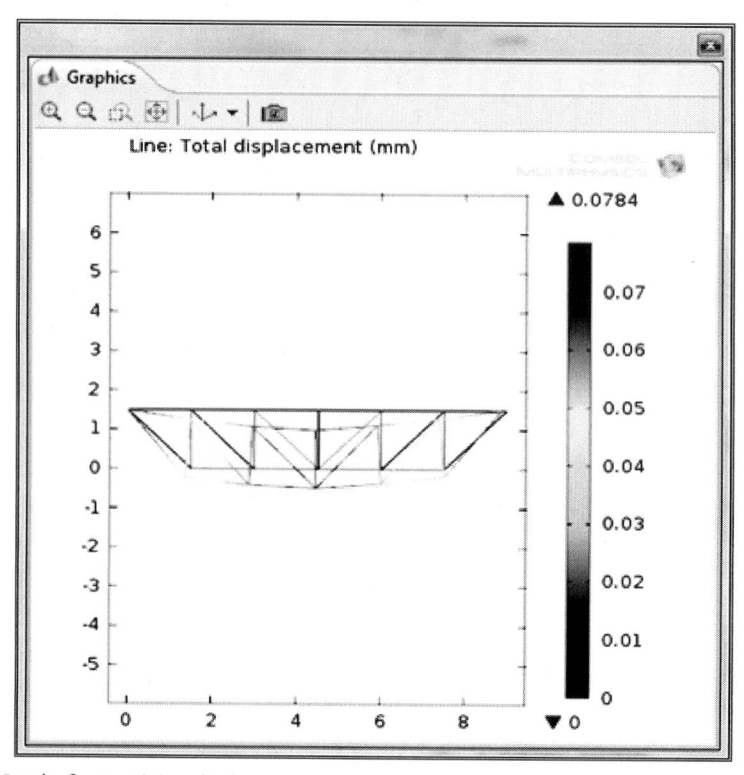

FIGURE 4.53 Results for truss joints displacements.

16. Right-click on the Example 4.truss2D.mph (*root*) node in the Model Builder window and select Add Study. In the Model Wizard window, select Eigenfrequency and click Finish . A new Study node will appear in the Model Builder window. Rename it **Study 2-Eigenfrequency w/o load.** Also rename its corresponding solver to **Solver 2-Eigenfrequency** w/o load. Right-click on the Study 2-Eigenfrequency w/o load node and select Compute .

17. Default results show the first eigenfrequency (133.35 Hz) and the corresponding displacements for modal shapes, as shown in Figure 4.54. Change the units for displacements to **mm.** To show the results for any one of the other five eigenfrequencies, simply select one from the list in the Data section and click Plot . Results for first and last eigenfrequencies are shown below.

18. To calculate the eigenfrequencies with loads, right-click on Example 4.truss2D.mph(*root*) node in the Model Builder window and select

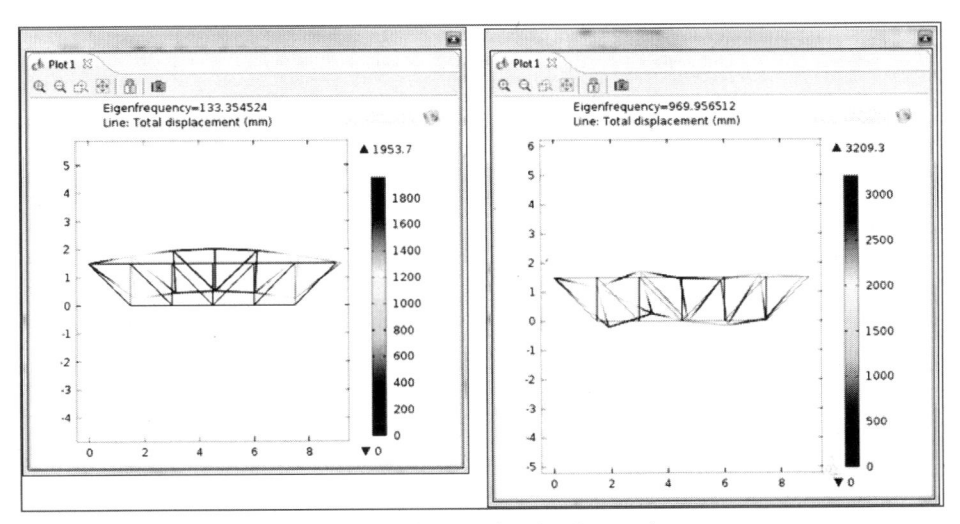

FIGURE 4.54 Results for truss displacement for two of its eigenfrequencies.

Add Study. In the Model Wizard window, select Prestressed Analysis, Eigenfrequency and click Finish. A new Study node will appear in the Model Builder window. Rename it **Study 3-Eigenfrequency with load.** Rename its corresponding solver to **Solver 3-Eigenfrequency** with load. Right-click on Study 3-Eigenfrequency with load node and select Compute. Notice that this study also adds a Stationary calculation.

19. The default results for first eigenfrequency (133.35 Hz) with corresponding displacements show up in the Graphics window. Since the difference between eigenfrequencies with and without loads is very small, we can ignore the effect of loading for calculating natural frequency of the truss.

    The first dynamic excitation of this truss (fundamental modal shape) happens at the frequency of 133.35 Hz. In other words, if a dynamic load with the same frequency is applied to this truss then resonance will happen and the truss will exhibit displacements with very large values. It would be useful to study the behavior of this truss for a range of harmonics, or applied loads with a range of frequencies including the fundamental natural one. We will do this in the following steps.

20. Right-click on the Example 4.truss2D.mph(*root*) node in the Model Builder window and select Add Study. In the Model Wizard window, select Frequency Domain and click Finish. A new Study

node will appear in the Model Builder window. Rename it **Study 4-frequency domain** and expand it. Click on the Step1: Frequency Domain node. In the corresponding window, enter **20, 60, 99, range(100,3,140), 150, 200, 300, 320** for the Frequencies in the Study Settings section. These are harmonics for a series of applied dynamic loads. We have a higher resolution around 133.35 Hz to capture the resonance. Also rename its corresponding solver to **Solver 4-frequency domain**. Right-click on the Study 4-frequency domain node and select Compute ≡.

21. The default results for member normal forces and stresses will appear in the Graphics window, as shown in Figure 4.55, for the last frequency at 320 Hz. Reults for harmonics can be shown by selecting the desired value from the list of Parameter values (*freq*) in the Plot window.

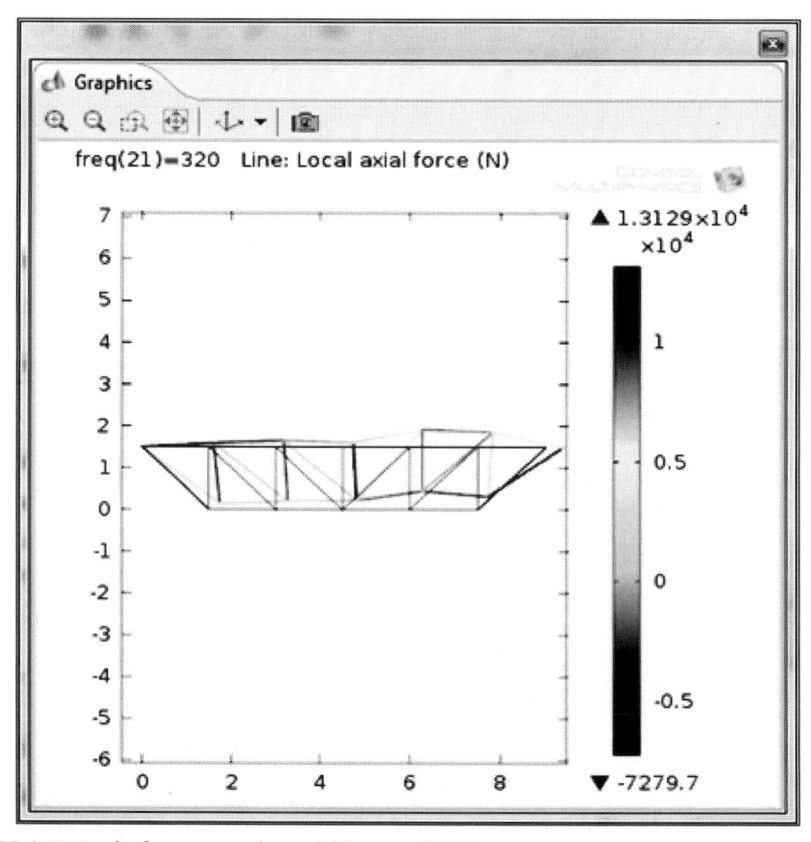

FIGURE 4.55 Results for truss members axial forces at 320 Hz.

22. It would be useful to draw the displacement of a point, say at the middle of the truss (node D) for a range of harmonics. Right-click on the Results node in the Model Builder window and select 1D Plot Group. A new 1D Plot Group 1 will appear in the model tree. Rename it disp. For harmonics. Right-click on this node and select Point Graph. In the corresponding window, select nodes representing D, H, and F from the Graphics window and add them to the Selection list. Change the unit to **mm** under Unit. Click Plot. The results for displacements for the range of harmonics will appear in the Graphics window, as shown in Figure 4.56.

These results show the resonance close to 133 Hz. So far for this model we have not introduced damping to the vibration. In practice damping is the capacity of the structure to absorb and dissipate part of the energy applied to the structure. We add the damping either by adding extra dampers or it is provided by the material used as a result of intrinsic friction. In COMSOL, damping can be added to the model using Rayleigh Damping or Loss Factor Damping. By using an isotropic structural loss factor of 0.01, we introduce damping to the model.

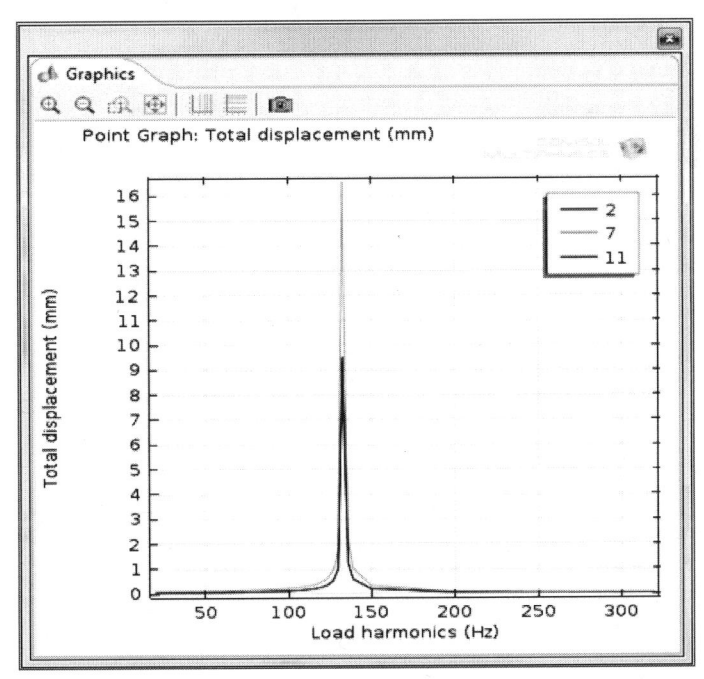

FIGURE 4.56 Total displacement at joints D, H, and F for different values of harmonics. The resonance is clearly shown.

23. Expand the Truss (*truss*) node in the Model Builder window, right-click on Linear Elastic Material 1, and select Damping. From the Damping window, locate Damping settings and select Isotropic loss factor from the list for Damping type. Select From material for Isotropic loss factor. Now we need to add the loss factor to the list of material properties. Expand Materials node and click on Structural steel (*mat1*). In the corresponding window, from the list under Material Contents, enter **0.01** for the value of eta_s, Isotropic structural loss factor, as shown in Figure 4.57.

24. To run the model, right-click on the Study 4-frequency domain node in the Model Builder and select Compute ≡. Wait for computations to finish.

25. Click on disp. For harmonics to see the results for damped values for displacements, as shown in Figure 4.58.

26. When damping is added to the model, COMSOL automatically considers it for eigenfrequency calculations, as well. In order to eliminate this, expand Study 2- Eigenfrequency w/o load node and click on Step1: Eigenfrequency. In the corresponding window, expand the Physics and Variables Selection and check the box for Modify physics tree and variables for study step. From the list, locate Damping 1 and click on it. Then click the Disable icon ⊘, as shown in Figure 4.59.

27. Repeat Step 25 for Study 3- Eigenfrequency with load.

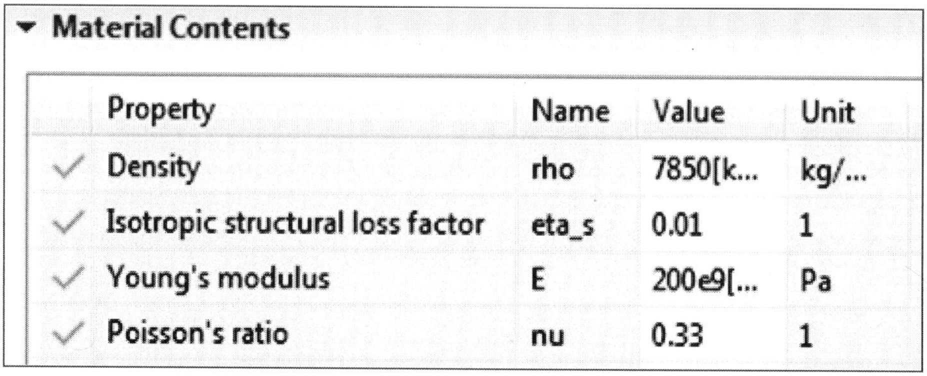

Property	Name	Value	Unit
✓ Density	rho	7850[k...	kg/...
✓ Isotropic structural loss factor	eta_s	0.01	1
✓ Young's modulus	E	200e9[...	Pa
✓ Poisson's ratio	nu	0.33	1

FIGURE 4.57 Window for adding damping as Isotropic loss factor.

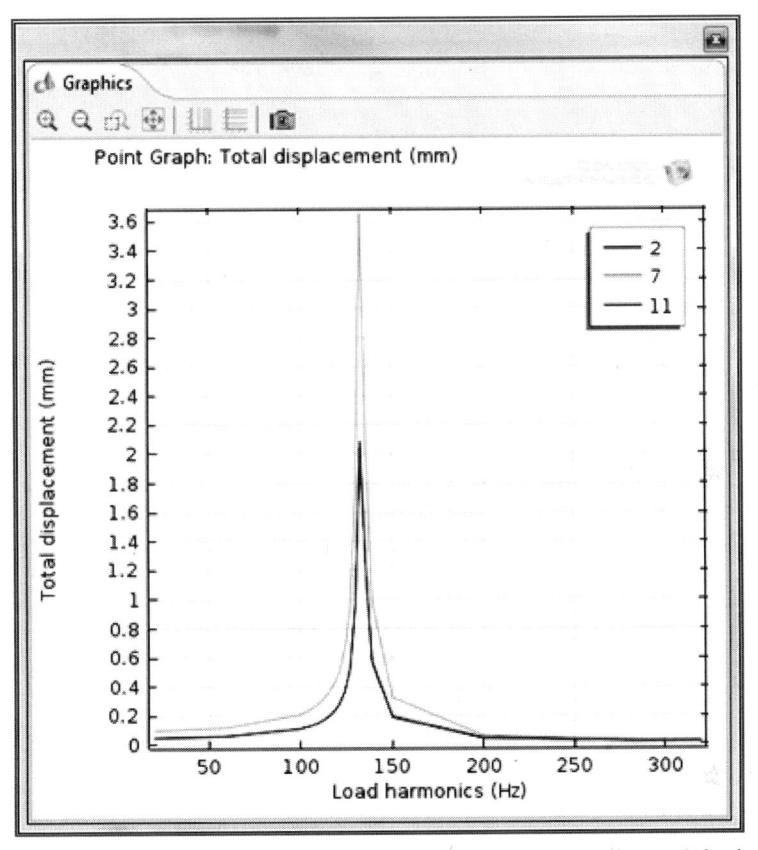

FIGURE 4.58 Total displacement at joint D with damping for different values of harmonic loads.

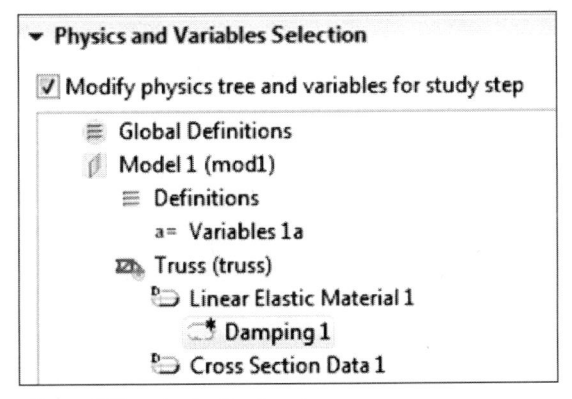

FIGURE 4.59 Window for modifying physics for damping.

Example 4.6: Static and Dynamic Analysis for a 3D Truss Tower

In this example, we use tools available in COMSOL to analyze a 3D structure under static load. We also calculate its natural frequencies. Shape and dimensions are given in Figure 4.60 and Table 4.1.

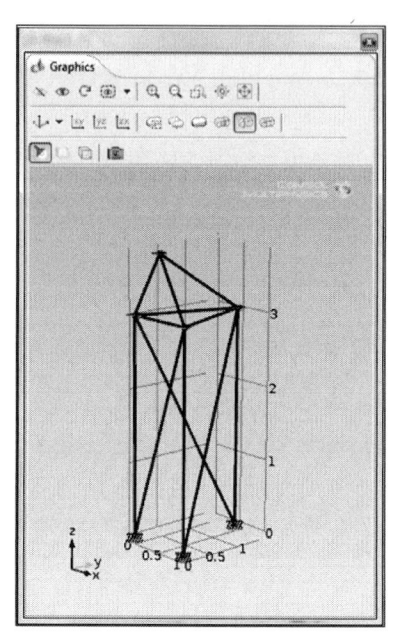

FIGURE 4.60 Geometry for the 3D truss.

Node	X(m)	Y(m)	Z(m)
A	0	0	0
B	1	0	0
C	0.5	$\sqrt{2}$	0
D	0	0	3
E	1	0	3
F	0.5	$\sqrt{2}$	3
G	0.5	0	$3+\sqrt{3}/2$

TABLE 4.1 Joint coordinates for the 3D truss.

Solution

1. Launch COMSOL and open a new file. Save it as **Example 4.6.**

2. In the Model Wizard window, select 3D button and click on Next ⇨.

3. From the Add Physics list, expand the Structural Mechanics node and from the list select Truss (*truss*), then add it to the Selected physics list by clicking on ✚. Then Click Next ⇨.

4. From the Select Study Type, select Stationary and click Finish 🏁.

   Now we draw the geometry of the truss in the Graphics window using CAD tools available in COMSOL.

5. Right-click on Geometry 1 and select More Primitives>Bezier Polygon. In the corresponding window, locate Polygon Segments and click Add Linear. Enter the coordinates for nodes A&D for member AD. Click Build Selected 🔳. Rename Bezier Polygon 1 to Bezier Polygon ad (b1). Right-click on Bezier Polygon ad (b1) and select Duplicate. In the corresponding window, enter the coordinates for nodes A&E. Similarly, draw the remaining truss members. Results are shown in Figure 4.61.

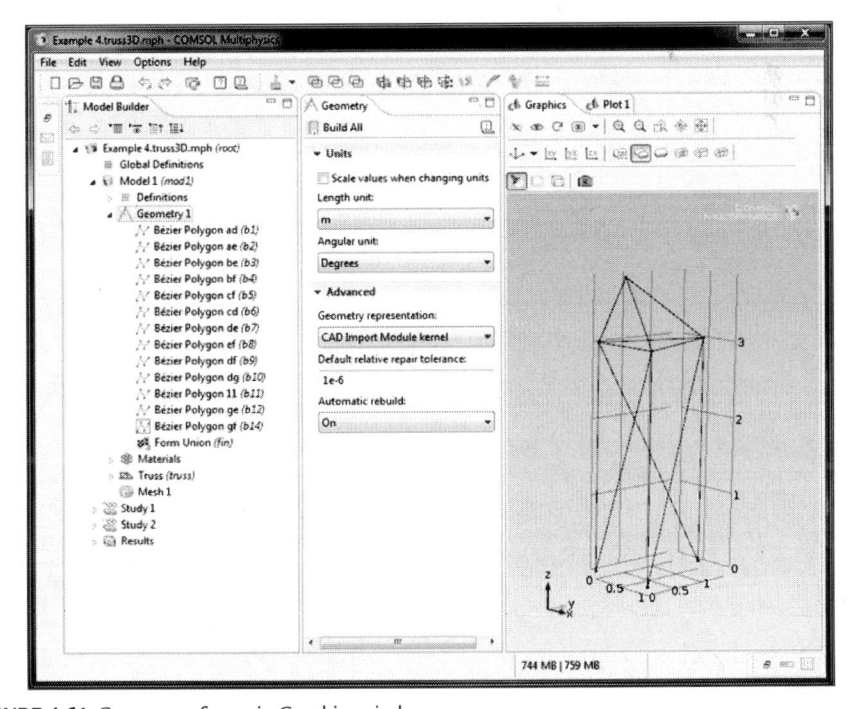

FIGURE 4.61 Geometry of truss in Graphics window.

6. Add materials to the model. Right-click on the Materials node in the Model Builder window and select Open Material Browser. In the Material window, as shown in Figure 4.62, expand Built-In and select Aluminum. Add this material to the model by clicking on ➕.

7. Right-click on Truss (truss) and select Cross Section Data. A new node will be added to the tree; rename it **L2x2x3/8 in**. In the corresponding window, enter **8.8E-4** in the space provided for A, under Area.

8. To define the boundary conditions (support types) and point loads, right-click on Truss (*truss*) and select Pinned. In the corresponding window, select nodes representing A, B, and C from the Graphics window and add them to the Selection list. Right-click again at the Truss (*truss*) node and select point Load. In the corresponding window, select the node that represents D and add it to the Selection list. In the same window, enter 150 for the value of **y** in the Force section.

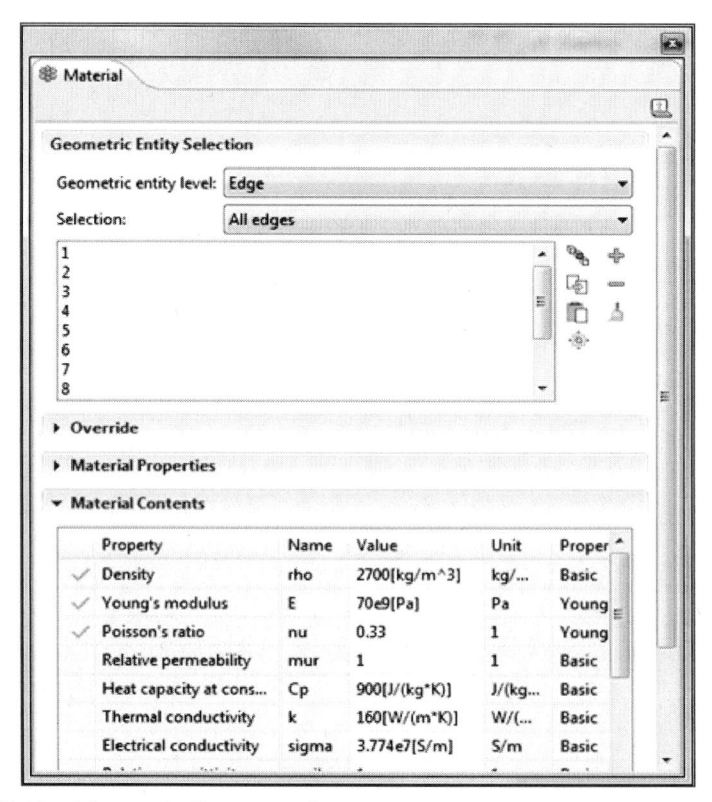

FIGURE 4.62 Material properties for truss members.

Similarly, add three more point loads as follows: for point G, force components are (0, 350, −150); for point E, force components are (0, 200, 0); and for point F, force components are (−300, 0, 0).

9. To build a mesh, click on Mesh 1 node in the Model Builder window, and in the Mesh window select Coarse from the list for Element size. Click Build All . A total of 40 elements are built.

10. Right-click on Study 1 and select Compute . Wait for computations to finish.

11. The default result will appear in the Graphics window showing truss members forces and stresses, as shown in Figure 4.63.

Although from these results we can read normal forces of the truss members using the legend, it would be useful to have the exact numerical values. We will define some variables for normal forces and extract their values from the results database.

12. Right-click on the Definitions node under Model 1 in the Model Builder window, and select Model Couplings>Average. From the Graphics window, select the member that represents AD and add it to the Selection list in Average window. Rename the node Average 1 (*aveop1*) to Average 1 (*aveop_ad*), as shown in Figure 4.64.

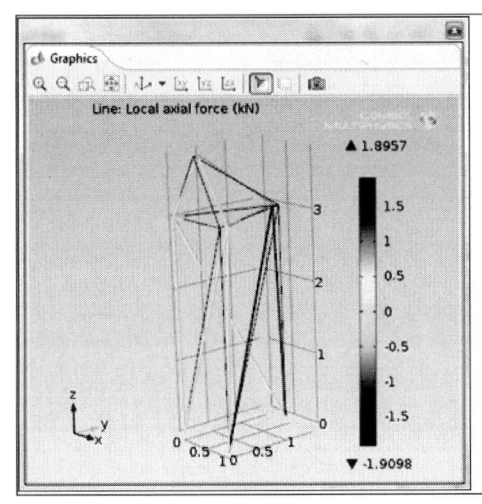

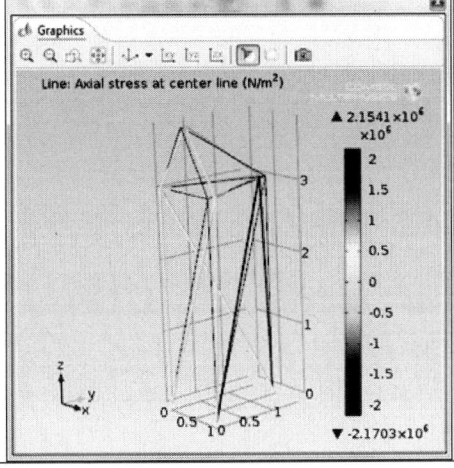

FIGURE 4.63 Results showing axial forces and stresses for truss members.

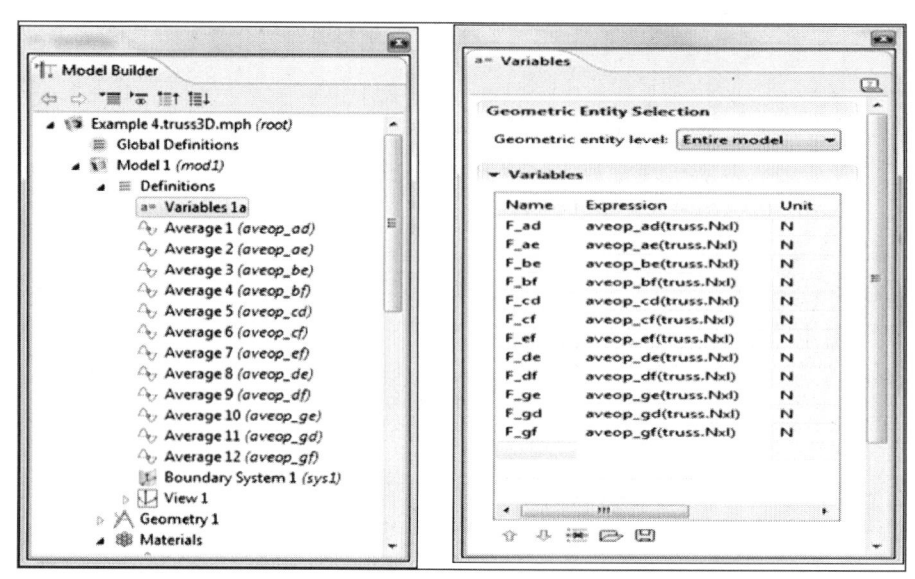

FIGURE 4.64 Variables and their definitions.

13. Repeat Step 12 for all remaining members of the truss.

14. Right-click on the Definitions node and select Variables. In the corresponding window, add the following list, as shown in Figure 4.64.

15. To see the numerical results for members, such as member BF, right-click on Derived Values located under Results node and select Global Evaluation. In the corresponding window, as shown in Figure 4.65, enter the following data and click Evaluate. The internal force for member BF would appear in the Results window, F_bf =0.65223 kN. Similarly, other members' forces can be extracted.

16. Now we would like to calculate the natural frequencies of the space truss. Right-click on Example 4.6.mph (root) node in the Model Builder and select Add Study. In the corresponding window, select Eigenfrequency from the list and click Finish.

17. A new node Study 2 will appear in the model tree. Click on Step 1: Eigenfrequency (located under Study 2 node) and in the corresponding window enter 6 for Desired number of eigenfrequencies (if needed, this number can be modified). This will set the model to calculate the first six eigenfrequenies. Another useful tool is Search for eigenfrequencies around:, which can calculate eigenfrequencies close to any desired value, if required.

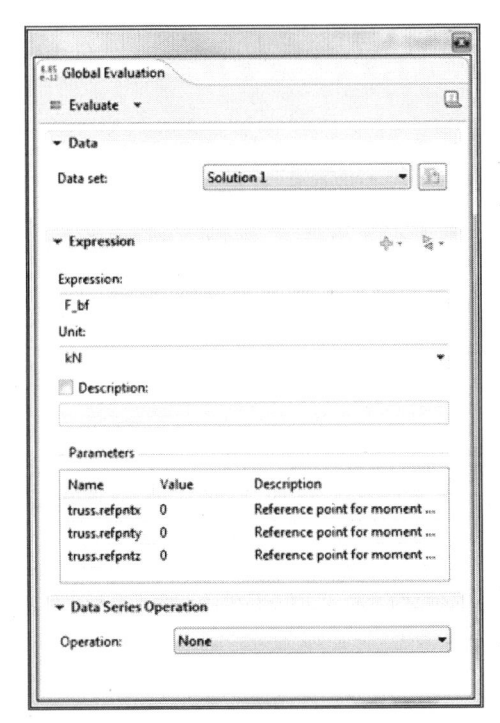

FIGURE 4.65 Global Evaluation of member BF axial force.

18. To run the model, right-click on Study 2 and select Compute ≡. Wait for the computations to finish.

19. The default result for natural frequency (first eigenvalue 32.88 HZ) will appear in the Graphics window. Click on the Mode Shape (*truss*) node and expand Window Settings section located in the corresponding window. From the list for Plot window, select Plot 1. This will create a new Graphics window for showing the results. Results for other values of eigenfrequencies can be plotted by choosing the desired value from the list. The results for first and last eigenfrequencies are shown in Figure 4.66.

20. To animate the modes of vibration for different eigenvalues, right-click on the Export node in the Model Builder window and select Player. In the Player window, modify the data as shown in Figure 4.67 and click Generate Frames ▦. Right-click on Player 1 and select Play ▷. To stop the animation, click on the Stop ▪ button located on the toolbar in the Graphics window.

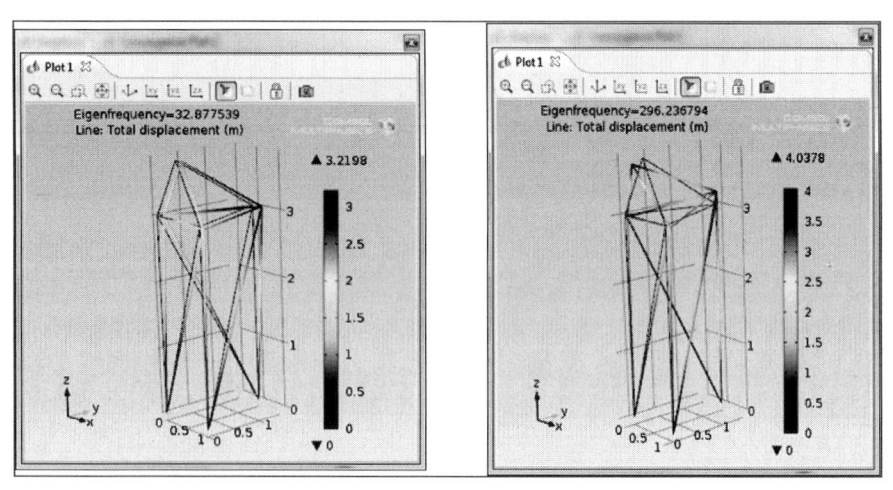

FIGURE 4.66 Results showing total displacements for two eigenfrequencies.

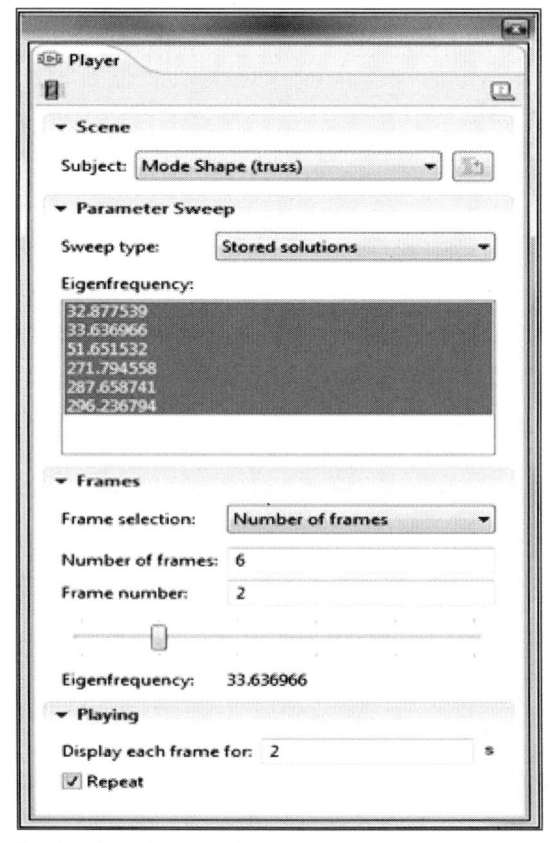

FIGURE 4.67 Results showing Player data entry for animation.

SECTION 4.2: DYNAMIC ANALYSIS AND MODELS OF INTERNAL FLOWS: LAMINAR AND TURBULENT

In this section, we use COMSOL modules to model some examples in fluid mechanics. Models include dynamic, parametric study, swirling, and moving boundary conditions of two- and three-dimensional flows, but mainly internal flows. Modeling and analysis of fluids flow is more complex than that for linear solid mechanics. This is mainly because of nonlinearity of governing equations (i.e., Navier-Stokes), as given below for an incompressible fluid. Fluid flow (u,v,w) are velocity vector components, p pressure, and (F_x, F_y, F_z) component of body force. ρ and μ are density and dynamic viscosity of the fluid, respectively.

$$\rho(u\frac{\partial u}{\partial x}+v\frac{\partial u}{\partial y}+w\frac{\partial u}{\partial z})+\frac{\partial p}{\partial x}=\mu[\frac{\partial^2 u}{\partial x^2}+\frac{\partial^2 u}{\partial y^2}+\frac{\partial^2 u}{\partial z^2}]+F_x$$

$$\rho(u\frac{\partial v}{\partial x}+v\frac{\partial v}{\partial y}+w\frac{\partial v}{\partial z})+\frac{\partial p}{\partial y}=\mu[\frac{\partial^2 v}{\partial x^2}+\frac{\partial^2 v}{\partial y^2}+\frac{\partial^2 v}{\partial z^2}]+F_y$$

$$\rho(u\frac{\partial w}{\partial x}+v\frac{\partial w}{\partial y}+w\frac{\partial w}{\partial z})+\frac{\partial p}{\partial z}=\mu[\frac{\partial^2 w}{\partial x^2}+\frac{\partial^2 w}{\partial y^2}+\frac{\partial^2 w}{\partial z^2}]+F_z$$

$$\frac{\partial u}{\partial x}+\frac{\partial v}{\partial y}+\frac{\partial w}{\partial z}=0$$

In engineering and industry we encounter many problems that require analysis and modeling of flow of a fluid and, in most cases, around or inside complex geometries. A fluid flow is categorized as laminar or turbulent when the Reynolds number is small or large, respectively, as compared to unity. The Reynolds number is a dimensionless number that measures inertia forces against viscous forces (in general) applied to a fluid point or particle. When the Reynolds number is large, flow becomes unstable and additional equations are needed to analyze the resulting turbulent flow. This adds to the complexity of flow analysis. Whereas for low-Reynolds flows, the inertia force can be neglected and the governing equation is a linear version of the Navier-Stokes equation or so-called Stokes equation.

Our main objective is to provide, for users and readers, some solved examples that can be used directly or lead to further solutions of similar or more complex flows using COMSOL. It is assumed that readers are familiar with relevant engineering principles and governing equations of fluid mechanics. In each example, we provide brief explanations of physics involved along with governing equations and phenomena, as applicable. It is recommended that readers cover Chapters 1 through 3 before attempting examples in this section.

Example 4.7: Axisymmetric Flow in a Nozzle: Simplified Water-jet

Axisymmetric flows may exist in many industrial and engineering problems. By definition an axisymmetric flow (or geometry) exists when the flow variables do not vary about the axis of symmetry involved. For example, in a tube for a fully developed flow the velocity along the axis of the cylinder at a given radius does not change with respect to the angular dimension. In other words, if an observer at a given radius moves around the axis of the tube then s/he will not observe any changes in the value of fluid velocity. The realization of axisymmetric flow, if it exists, is very important since it could reduce a 3D flow analysis to a 2D one.

In this example, we model flow in a nozzle that has an axis of symmetry. The cross-section and dimensions of the nozzle in the r-z plane are shown in Figure 4.68.

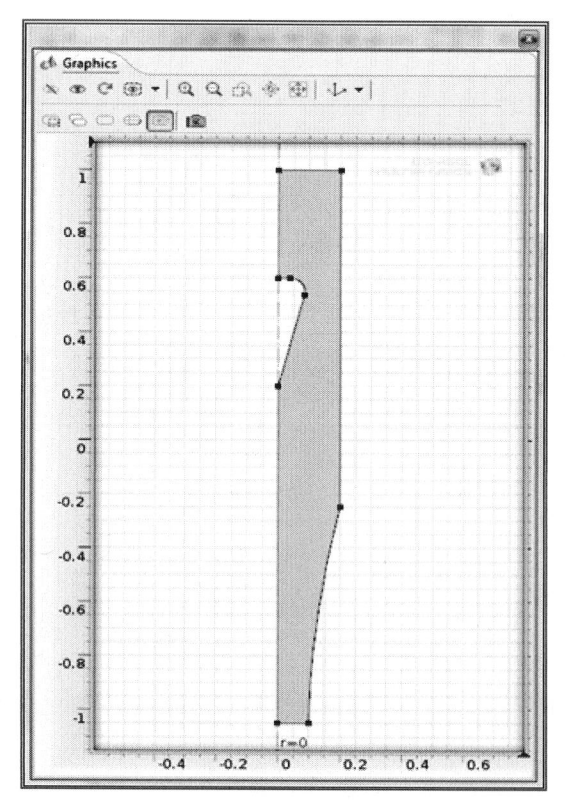

FIGURE 4.68 Geometry and dimensions of the Nozzle cross-section in (r-z) plane.

Solution

1. Launch COMSOL and open a new file. Save the file as Example 4.7.

2. Select 2D axisymmetric from the list (in the Select Space Dimension section) in the Model Wizard window and click ➪ Next.

3. In the Add Physics window, open the list under Fluid Flow and select Single-Phase Flow>Laminar Flow. Add this selection to the Selected physics list by right-clicking (or click on the ✚). Then click on ➪ to go to the next step. See Figure 4.69.

4. In the Select Study Type node, click on Stationary and then click on 🏁 Finish.

5. Draw the geometry (or import a CAD file) of the nozzle cross-section as shown in Figure 4.71. Right-click on the Geometry node in the

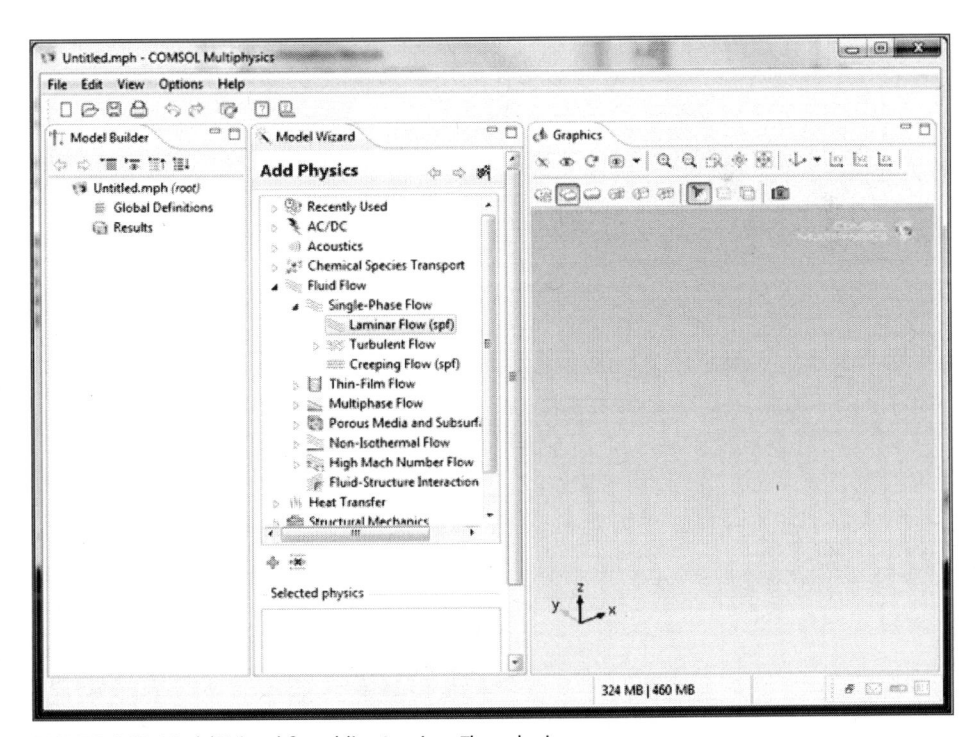

FIGURE 4.69 Model Wizard for adding Laminar Flow physics.

Model Builder and select Bezier Polygon from the list. In the Bezier Polygon window, click on the Add Linear button and enter the following data:

	r:	z:	
1	0	1	m
2	0.2	1	m

6. Similarly, add another line segment by clicking on the Add Linear button and enter the following data:

	r:	z:	
1	0.2	1	m
2	0.2	-0.25	m

7. Add a quadratic line by clicking on the Add Quadratic button and enter the following data:

	r:	z:	
1	0.2	-0.25	m
2	0.1	-0.65	m
3	0.1	-1.05	m

8. Add two more lines with following data, and then click the Build Selected icon.

	r:	z:	
1	0	-1.05	m
2	0	1	m

	r:	z:	
1	0.1	-1.05	m
2	0	-1.05	m

See Figure 4.70.

9. Right-click on the Geometry node in the Model Builder window and select Bezier Polygon. In the Bezier Polygon window, click on Add Linear to create three line segments with following data:

	r:	z:	
1	0	0.6	m
2	0.1	0.6	m

	r:	z:	
1	0.1	0.6	m
2	0	0.2	m

	r:	z:	
1	0	0.2	m
2	0	0.6	m

10. Click on the Build Selected icon to create the triangle. To remove the triangular shape, click anywhere in the Graphics window and press the CTRL+A keys. The geometry will change to a red color. Click on the Difference icon located in the toolbar. See Figure 4.71.

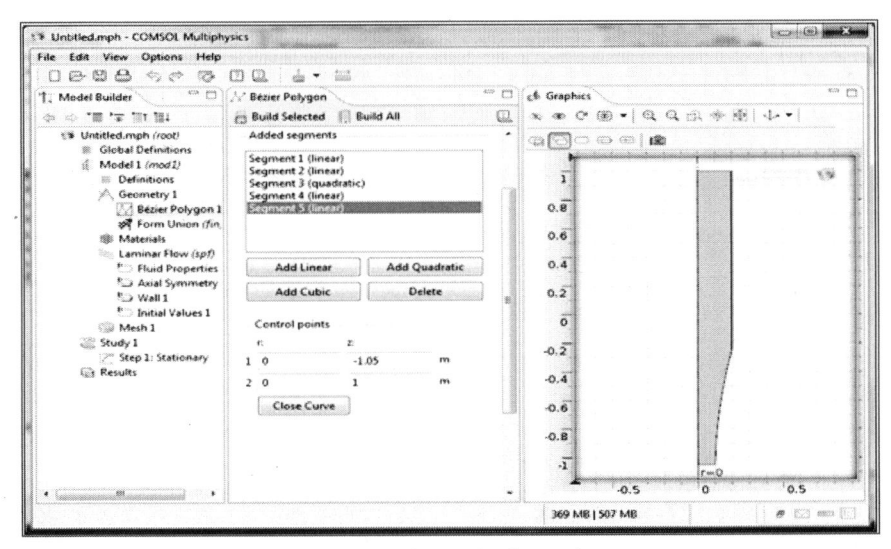

FIGURE 4.70 Geometry and line segments coordinates for the nozzle.

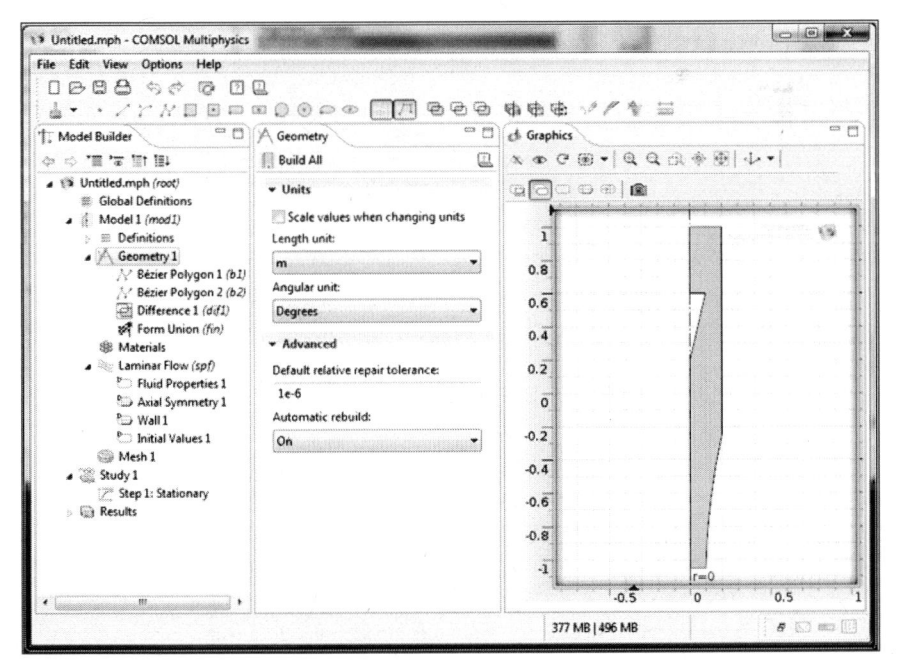

FIGURE 4.71 Geometry cross-section with the cut.

11. Add a fillet to the sharp corner of the triangular cut. Right-click on the Geometry node in the Model Builder and select Fillet. The Fillet window will open. In this window, add the vertex of the sharp corner of the triangle (click on the vertex and then right-click) to the list under vertices to fillet. For Radius enter 0.03 (unit should read m) and click Build Selected icon . At this point the geometry of the nozzle cross-section is complete, as shown in Figure 4.72.

12. To add material, right-click on the Materials node in the Model Builder and select Open Material Browser. In the Material Browser window, expand Liquid and Gases>Liquids and select Water from the list by right-clicking on it and selecting Add Material to Model. The properties of water are given as functions of its temperature as shown under Material Contents section (expand this section if needed), as shown in Figure 4.73.

13. This model is for an isothermal flow in the nozzle, but for calculating the properties of water the value of temperature is required. Right-click on the Global Definitions in the Model Builder window and select Parameters. In the Parameters window, add the data as shown in Figure 4.74 (alternatively, the value of temperature can be added in the Model Inputs section). Click on the Fluid Properties node to open the Fluid Properties window. We also add water speed at the inlet as a parameter.

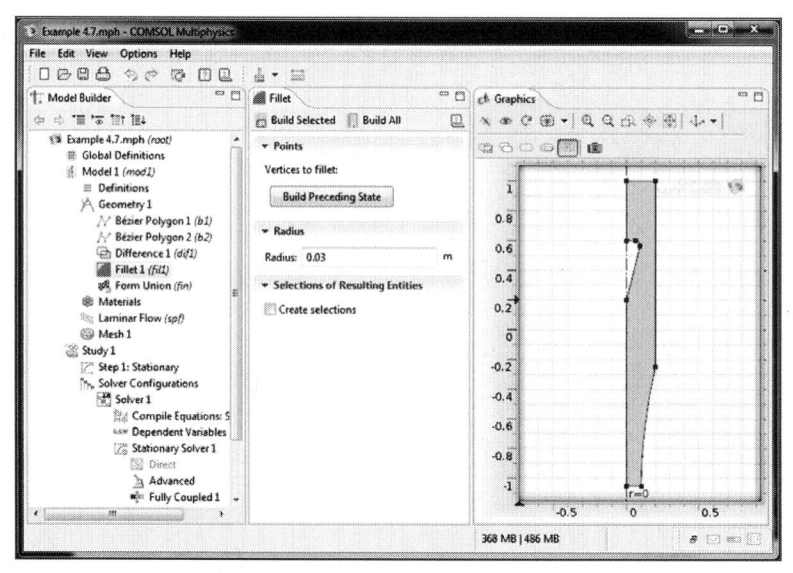

FIGURE 4.72 Geometry cross-section with the cut and fillets.

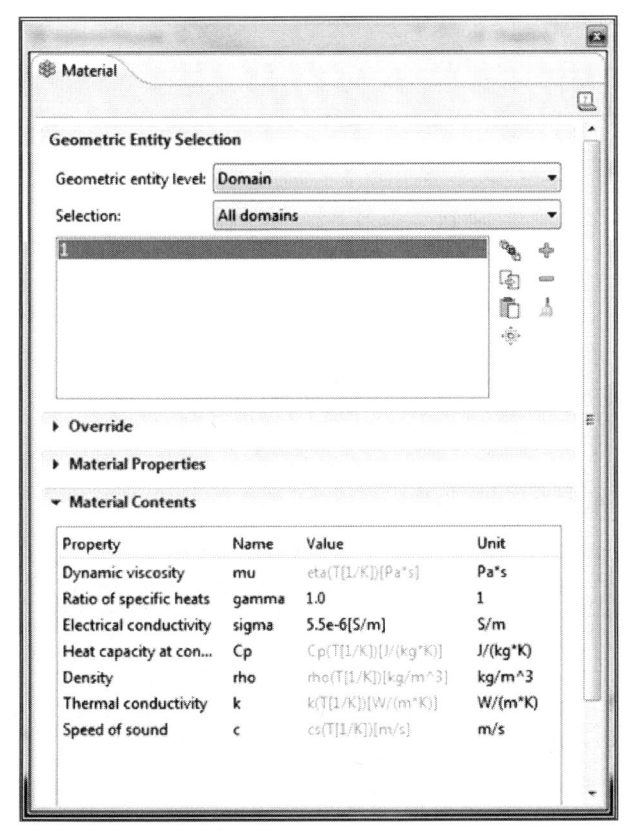

FIGURE 4.73 Material properties for nozzle.

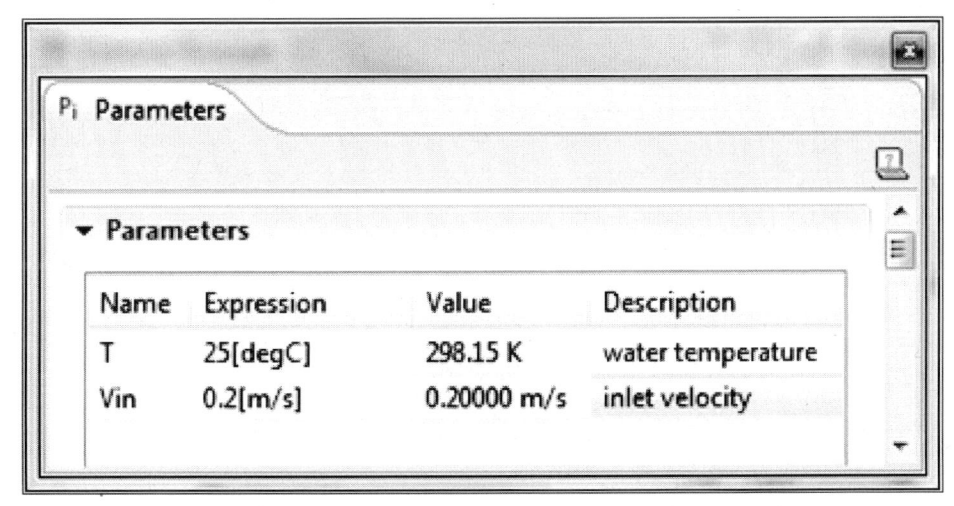

FIGURE 4.74 Parameters data entry.

14. To define the boundary conditions, right-click on the Laminar Flow (*spf*) node in the Model Builder window and select Inlet. In the Inlet window, add the edge at the top of geometry (z = 1) and right-click. In the Velocity section, enter **Vin** for U_0 (make sure that Normal inflow velocity is checked). This will assign the water velocity to the value set in the Parameters (**Vin**) entering into the nozzle from the top. Similarly, add Outlet boundary condition at the exit (the bottom edge) and set the Pressure value P_0 equal to 0. Double-check the Axial Symmetry 1 (automatically created) to have the vertical edges at r = 0 as axis of symmetry. See Figure 4.75.

15. To create a mesh, click on the Mesh 1 node in the Model Builder window. In the Mesh window under Sequence type, select User-controlled mesh. Click on the Size node under Mesh 1 in the Model Builder window, and in the Size window select Extremely coarse from the Predefined list. Expand the list under Element Size Parameters to see the specifications of created mesh. See Figure 4.76.

16. To set up a parametric study based on inlet velocity, right-click on the Study 1 node and select Parametric Sweep. In the Parametric Sweep

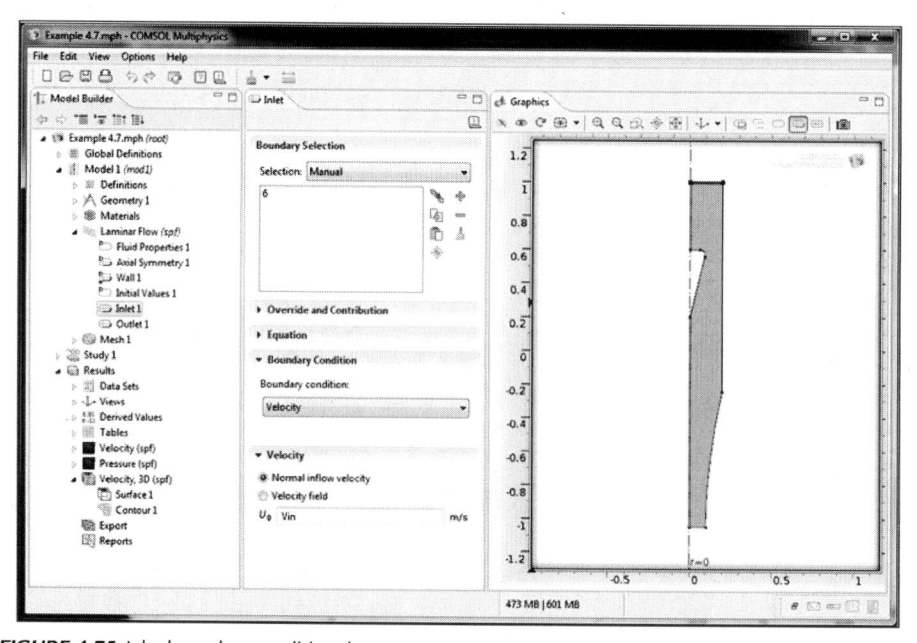

FIGURE 4.75 Inlet boundary condition data.

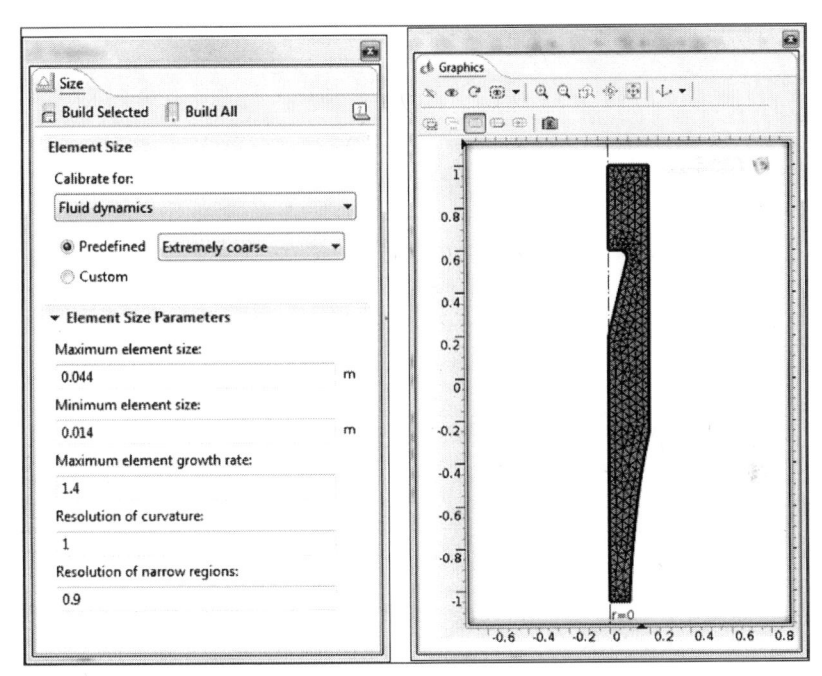

FIGURE 4.76 Mesh data and the final mesh.

window under the Sweep type table, click on the plus sign ✚ and open the list under Parameter names that appears in the table. From the list select **Vin** (inlet velocity) and type 0.5, 1, 3, 5 for Parameter value list, as shown in Figure 4.77. This sets the value of inlet velocity for a sequence of runs for the model.

17. To run the model, right-click on Study 1 node in the Model Builder window and select Compute .

18. To show the results in 3D (recall that this model is axisymmetric), right-click on Velocity 3D (spf) node and select Surface. Repeat this operation and select Contour. Click on Contour 1 node and in the Contour window type **p** for the Expression. Click on the Plot icon . The velocity mapped by pressure contours will appear in the Graphics window. To see these values for different input velocity values, simply click on Velocity 3D (spf) node, and in the corresponding window select the desired value for Parameter value (**Vin**), under the Data section. Results for **Vin=5 m/s** are shown in Figure 4.78.

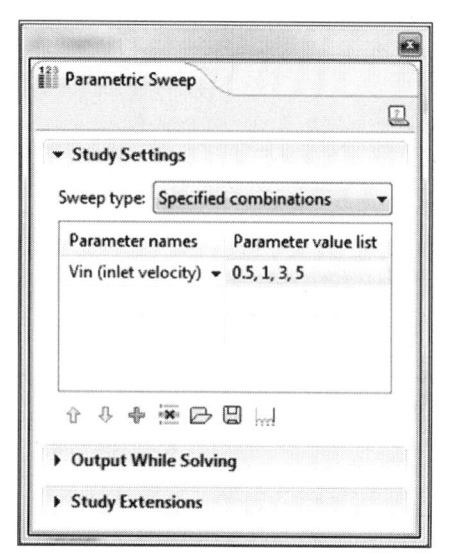

FIGURE 4.77 Parametric Sweep values for values of inlet velocity.

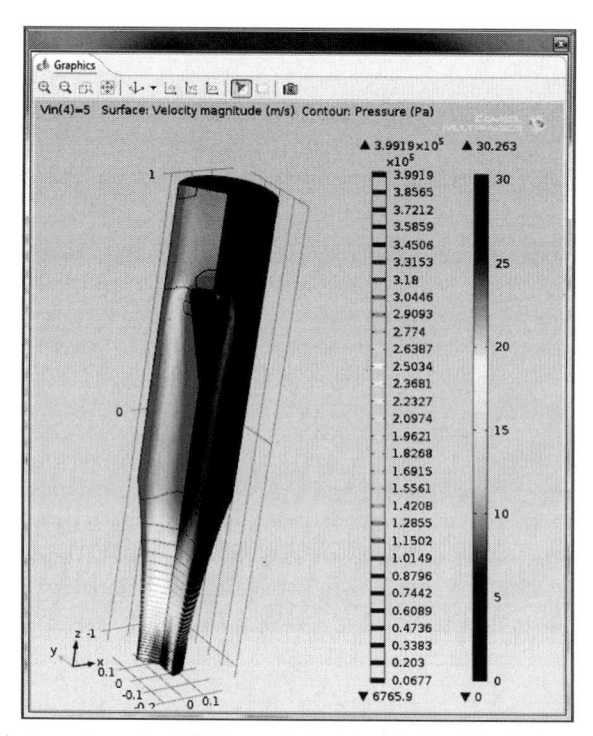

FIGURE 4.78 Results for surface velocity in the nozzle for inlet velocity of 5 m/s.

Example 4.8: Swirl Flow Around a Rotating Disk: Laminar Flow

In this example, we rebuild in COMSOL version 4.3 a model from the COMSOL Model Library (CFD_Module/Single-Phase_Tutorials/rotating_disk.)[2] This model is a 3D flow in an axisymmetric geometry. We use this model as another example to demonstrate how to model a turbulence flow.

Flow around a rotating disk happens in many industrial processes and mechanical machines, such as high-speed fly wheels for energy storage and mixers. Calculation of stress field in the fluid and consequently on the rotating disk itself is required for design of the system for both low-speed and high-angular velocities cases.

The 3D geometry of the container and rotating disk is shown schematically in Figure 4.79. When the disk rotates, the flow in the tank will eventually reach a steady-state. Since the geometry is axisymmetric, we use a 2D axisymmetric model. However, the flow is three-dimensional and the fluid velocity vector has components in radial, rotational, and axial directions. The 2D axisymmetric geometry is also shown in Figure 4.79.

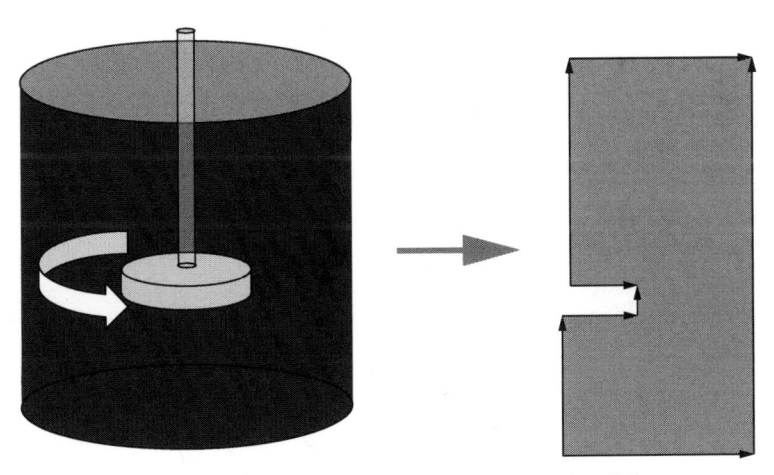

FIGURE 4.79 3D and 2D axisymmetric geometries for flow around a rotating disk.

---

[2] Model made using COMSOL Multiphysics® and is provided courtesy of COMSOL. COMSOL materials are provided "as is" without any representations or warranties of any kind including, but not limited to, any implied warranties of merchantability, fitness for a particular purpose, or noninfringement.

The governing equations are Navier-Stokes equations. The momentum transfer equations for a stationary, axisymmetric flow are written as:

$$\rho(u\frac{\partial u}{\partial r} - \frac{v^2}{r} + w\frac{\partial u}{\partial z}) + \frac{\partial p}{\partial r} = \mu[\frac{1}{r}\frac{\partial}{\partial r}(r\frac{\partial u}{\partial r}) - \frac{u}{r^2} + \frac{\partial^2 u}{\partial z^2}] + F_r$$

$$\rho(u\frac{\partial v}{\partial r} - \frac{uv}{r} + w\frac{\partial v}{\partial z}) = \mu[\frac{1}{r}\frac{\partial}{\partial r}(r\frac{\partial v}{\partial r}) - \frac{v}{r^2} + \frac{\partial^2 v}{\partial z^2}] + F_\varphi$$

$$\rho(u\frac{\partial w}{\partial r} + w\frac{\partial w}{\partial z}) + \frac{\partial p}{\partial z} = \mu[\frac{1}{r}\frac{\partial}{\partial r}(r\frac{\partial w}{\partial r}) + \frac{\partial^2 w}{\partial z^2}] + F_z$$

where u is the radial velocity, v the rotational velocity, and w the axial velocity, μ viscosity, and p pressure. The body volumetric forces (F_r, F_φ, F_z) are all equal to zero in this model.

Solution

1. Launch COMSOL and open a new file. Save the file as Example 4.8.

2. Select 2D axisymmetric from the list (in the Select Space Dimension section) in the Model Wizard window and click ➪ Next.

3. In the Add Physics window, open the list under Fluid Flow and select Single-Phase Flow>Laminar Flow. Add this selection to the Selected physics list by right-clicking (or click on the ✚). Then click on ➪ to go to the next step.

4. In the Select Study Type, click on Stationary and then click on 🏁 Finish.

5. To define the disk angular velocity as a parameter, right-click on the Global Definitions in the Model Builder window and select Parameters. In the Parameters window, add the data Name: **omega** and Expression: **0.25*pi[rad/s],** as shown in Figure 4.80.

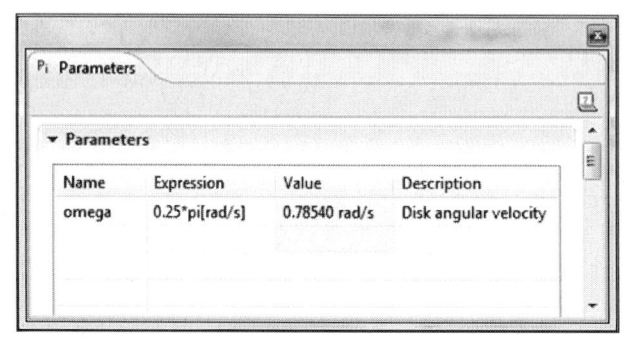

FIGURE 4.80 Defined parameters.

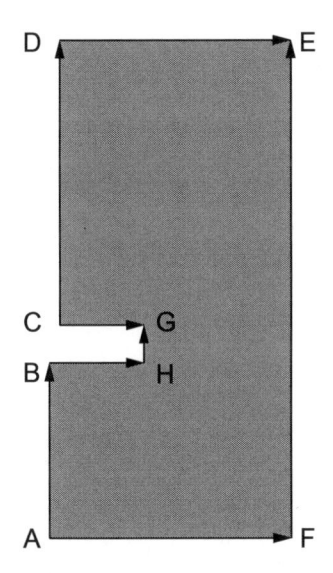

Edge	AB	BH	HG	GC	CD	DE	EF	FA
Length (m)	0.014	0.008	0.003	0.007	0.023	0.019	0.04	0.02

FIGURE 4.81 Axisymmetric geometry and dimensions for flow around a rotating disk.

6. Draw the 2D cross-section geometry with the dimensions as shown in Figure 4.81.

   Right-click on the Geometry 1 node in the Model Builder window and select Rectangle. In the Rectangle window (under the Size section), enter 0.02 for Width and 0.04 for Height. Under Position, select Corner for Base and enter 0 for r and 0 for z. Right-click on the Geometry 1 node in the Model Builder window and select Rectangle. In the Rectangle window (under the Size section), enter 0.008 for Width and 0.003 for Height. Under Position, select Corner for Base and enter 0 for r and 0.014 for z. Finally, right-click on the Geometry 1 node in the Model Builder window and select Rectangle. In the Rectangle window (under the Size section), enter 0.001 for Width and 0.023 for Height. Under Position, select Corner for Base and enter 0 for r and 0.017 for z. Click at any point in the Graphics window, press CTRL+A to highlight the geometry, then click on the Difference icon in the toolbar. Results are shown in Figure 4.82.

7. To add materials to the model, right-click Materials (under Model 1 in the Model Builder window) and select Material. The Material window will open. In this window under the Material Contents section, enter the data for density and dynamic viscosity, as shown in Figure 4.83.

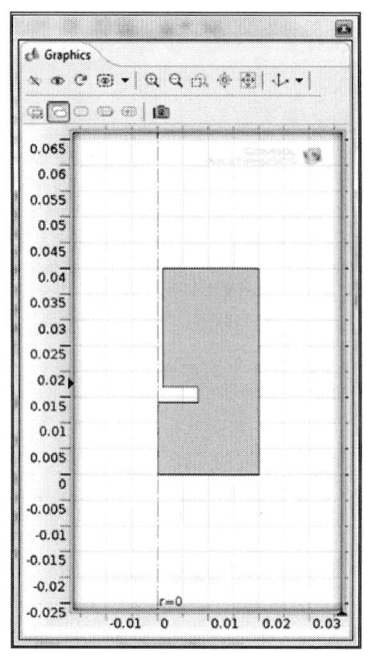

FIGURE 4.82 Axisymmetric geometry and dimensions.

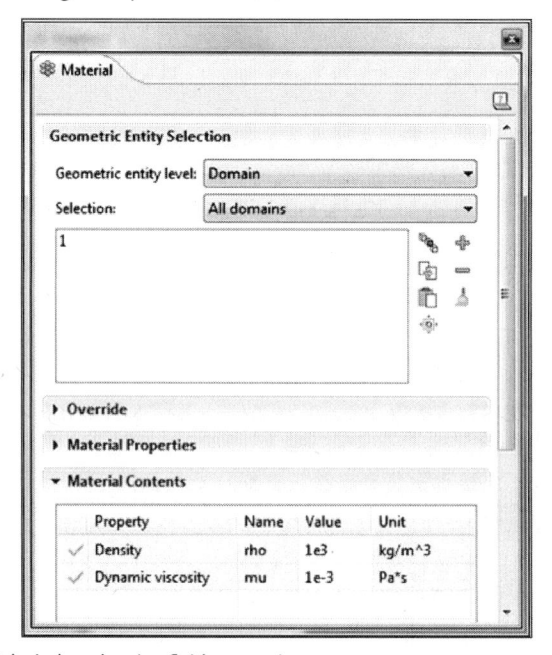

FIGURE 4.83 Material window showing fluid properties.

8. To define the details of the model physics and boundary conditions, click on the Laminar Flow in the Model Builder window. Expand the Physical Model section and select Incompressible flow from the list under Compressibility. Also check the box for Swirl flow, as shown in Figure 4.84.

9. Next we define the order of polynomial functions used for finite elements for velocity and pressure. The default is linear functions (P1 + P1); in this model we use a second order polynomial for velocity and a linear one for pressure (P2 + P1). This combination is recommended for Stokes and swirl flows (See Chapter 2 and the COMSOL Manual for *The Laminar Flow Interface-Discretization*). Click on the Show icon located in the toolbar of Model Builder window and select Discretization. Then click on the Laminar Flow 1(*spf*), and in the Laminar Flow window select P2 + P1 from the list under Discretization of fluids, as shown in Figure 4.85.

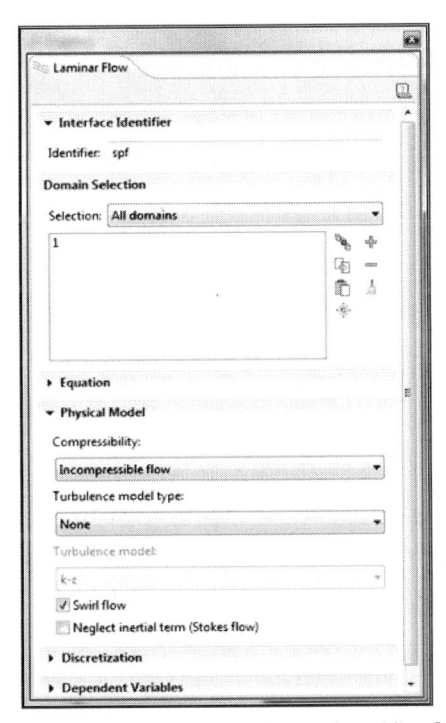

FIGURE 4.84 Laminar flow window showing fluid physics set up for swirling flow.

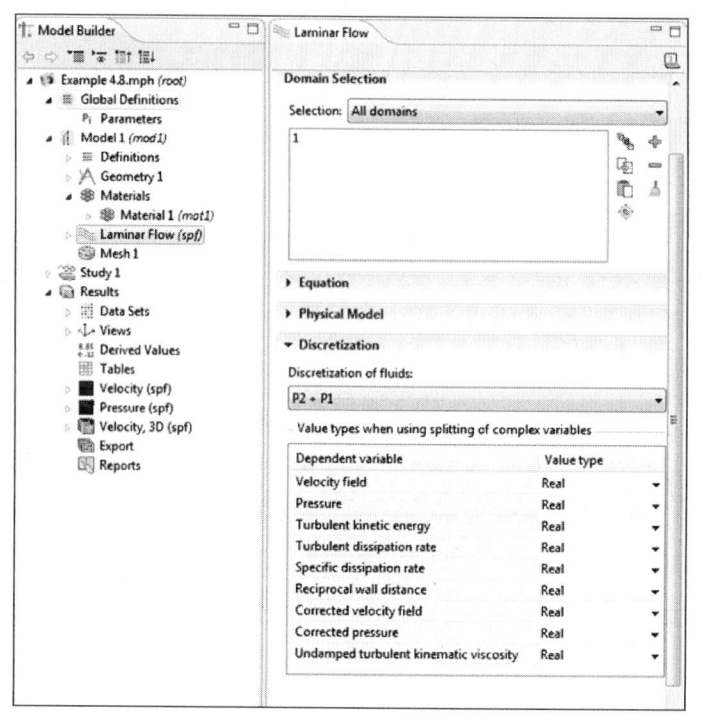

FIGURE 4.85 Laminar flow window showing discretization scheme setup.

Right-click on the Laminar Flow node and select Wall. The Wall window will open. In this window, select and add the boundaries (3, 4, 5, 7) related to the shaft and disk by clicking on these boundaries (one-by-one) and clicking on the plus icon ✚. In the same window under Boundary Condition, select Sliding wall from the list and type **omega*r** for V_w. This will set the rotational velocity of the disk as the boundary value for the fluid. See Figure 4.86.

For the boundary at the top surface of the fluid, we use symmetry boundary condition to allow for radial and rotational velocities and eliminate the velocity in the z-direction. Right-click on Laminar Flow (*spf*) node and select Symmetry from the list. In the Symmetry window, add the edge on the top (which reps the free-surface, edge #6) to the Selection by clicking on it in the Graphics window, and then right-click. To see the equations for this boundary condition, expand the Equation section.

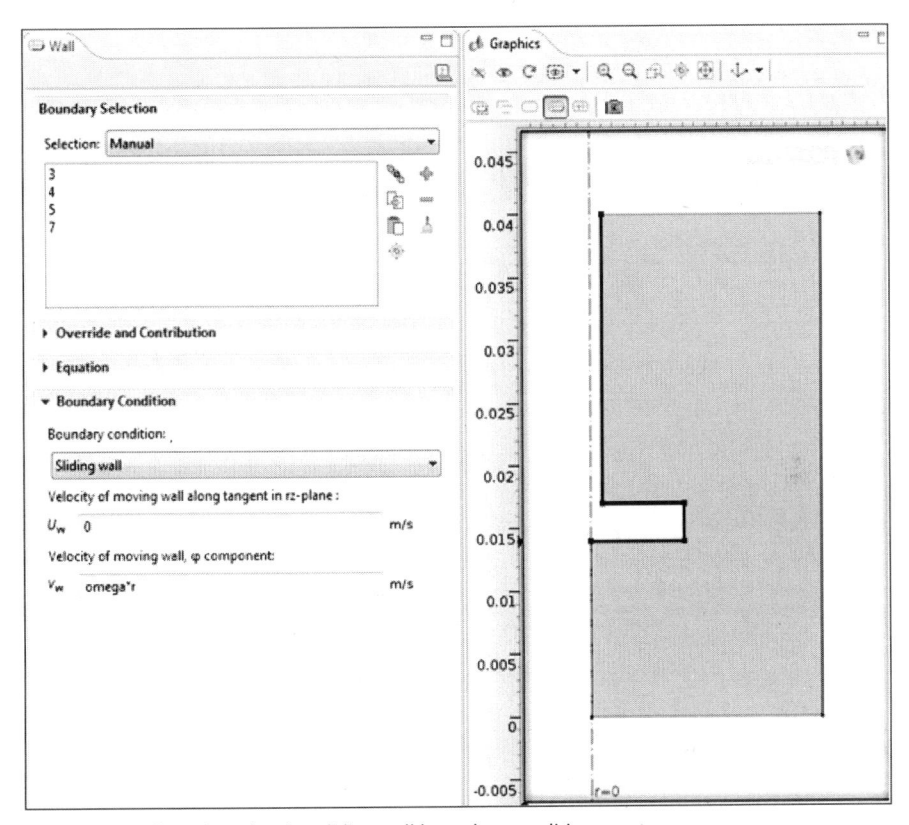

FIGURE 4.86 Wall window showing sliding wall boundary conditions set up.

Finally, we should set a reference point for the pressure since there is no output or exit point for the fluid. Right-click on the Laminar-Flow (*spf*) node and select Point>Pressure Point Constraint. In the corresponding window, add vertex at the top right corner of the geometry (#8) to the list by clicking on the vertex, and then right-click.

The boundary conditions applied to this model are listed in Figure 4.87.

Edge	AB	BH	HG	GC	CD	DE	EF	FA
Boundary conditions	Axial symmetry	Sliding wall	Sliding wall	Sliding wall	Sliding wall	Symmetry	No-slip wall	No-slip wall

FIGURE 4.87 Boundary conditions applied to the walls of the rotating disc.

10. To run the model for a series of values of disk angular velocities **omega,** expand the Study 1 node in the Model Builder window and click on Step 1: Stationary. In the Stationary window, locate the Study Extensions section and expand it. Check the Continuation box and click on the plus icon ✚. From the list, which appears under Continuation parameter, select omega (Disk angular velocity) and type **0.25*pi, 0.5*pi, 2*pi, 4*pi** under the Parameter value list, as shown in Figure 4.88.

11. Run the model by right-clicking on the Study 1 node and selecting Compute ≡. After calculations are finished, the default result will appear in the Graphics window.

12. To manipulate the results, click on Velocity (*spf*) and in the 2D Plot Group select 0.785398 from the list for Parameter value (omega), then click the Plot icon. Add the streamlines to the 2D plot for velocity. Right-click on Velocity (*spf*) and select Streamline. In the Streamline window, set the expressions as shown in Figure 4.89, then click Plot.

The model results for four values of omega are shown in Figure 4.90.

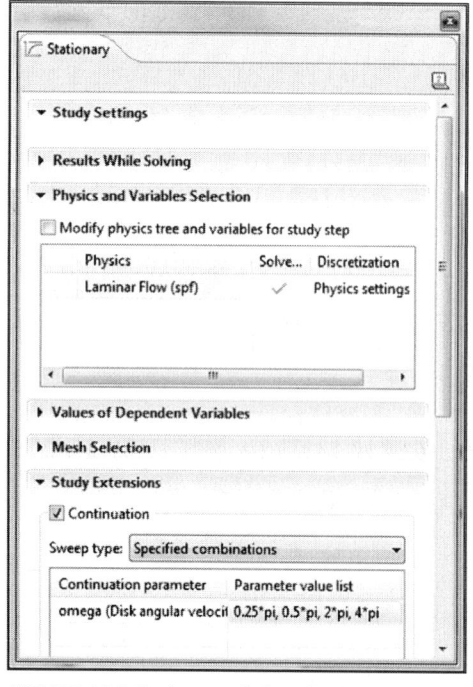

FIGURE 4.88 Stationary window showing parameter values set up for disk.

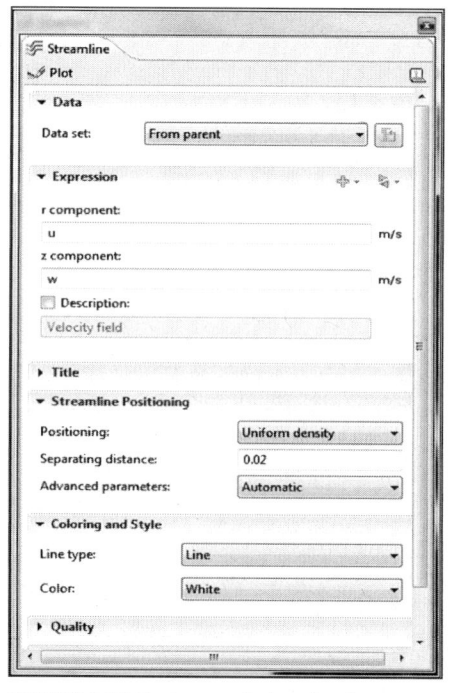

FIGURE 4.89 Stationary window showing parameter values set up for disk.

13. To see the 3D axisymmetric results, right-click on the Results node and select 3D Plot Group. The new Velocity, 3D (*spf*) will be created. Right-click on it and select Surface. In the Surface window, change the Unit to mm/s and click Plot. Click on Contour 1 and deselect the Color legend box, located under Coloring and Style. Zoom on the disk location to see the details. Results are shown in Figure 4.91.

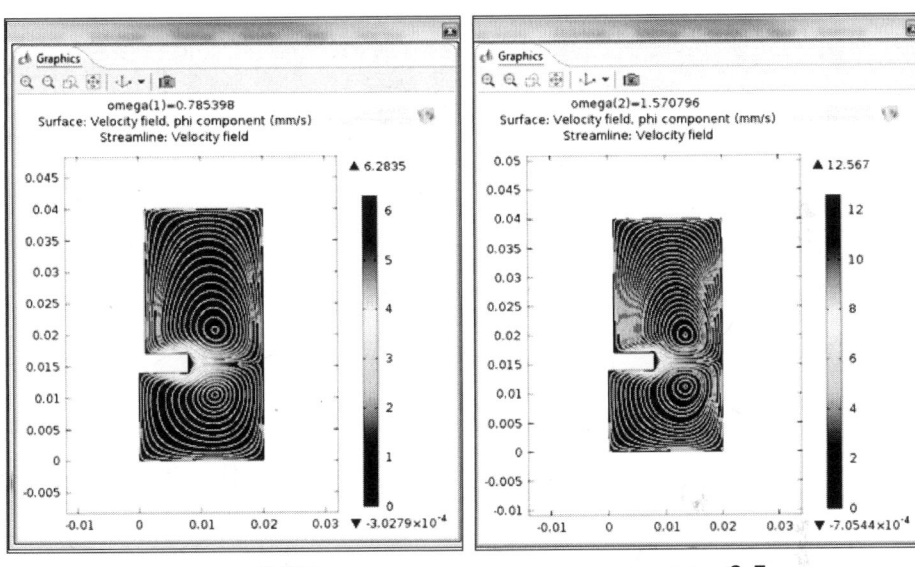

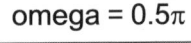

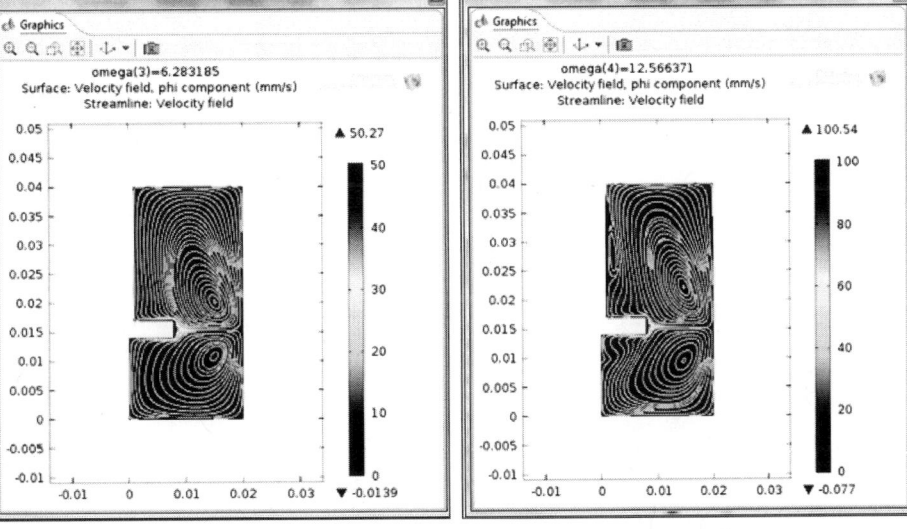

FIGURE 4.90 Results for 2D surface velocity for different values of omega.

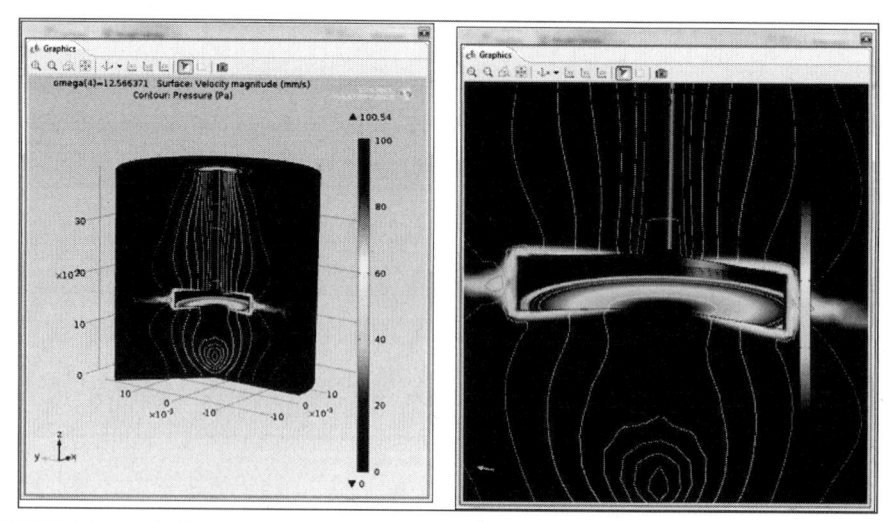

FIGURE 4.91 Results for 3D surface fluid velocity for a value of omega = 4π.

Example 4.9: Swirl Flow Around a Rotating Disk: Turbulent Flow

This example is an extension of Example 4.8 with turbulent flow modeling. For high values of the angular velocity of the rotating disk, we may have to consider turbulence flow. The value of the Reynolds number will be calculated to determine the state of the flow regime. We define the Reynolds number as $R_e = \dfrac{\Omega r^2}{\nu}$ where Ω is disk angular velocity, r is disk radius, and ν is fluid kinematic viscosity. COMSOL provides us with several turbulence models. For this example we use k-ε model, which is a RANS (Reynolds Averaged Navier Stokes equations) type model. The general approach in RANS is to assume decomposition of fluid velocity into a statistical average, such as the mean velocity, and a deviation from the mean. When governing N-S equations are averaged, the resulting equations contain additional stresses that are due to turbulence strains on the flow. These additional turbulence tresses are called Reynolds stresses and require extra equations for their solutions. One approach is to calculate the turbulence kinetic energy k and its dissipation rate ε (hence k-ε model) as additional equations. Turbulence modeling is still an R&D subject and advanced modeling techniques are still being investigated and validated. Interested readers are referred to text books (see Reference 4.5) and scientific journals for further study on this subject.

Solution

1. Launch COMSOL and open the Example 4.8 file. Save it as Example 4.9.

2. Click on the Show icon in the Model Builder toolbar and select Advanced Physics Options, as shown in Figure 4.92.

3. Click on the Laminar Flow (*spf*) node. In the Laminar Flow window, in the Physical Model section locate Turbulence model type and select RANS. From the list under Turbulence model select $k - \varepsilon$ (it is recommended to expand the Equation section and study the equations listed). Expand the Advanced Settings section and select the box for *Use pseudo time stepping for stationary equation form*, as shown in Figure 4.93. This is a common method in modeling; setting a pseudo time stepping for stationary flow models helps stability and convergence of the solution, especially for complex flows such as this one.

4. To build a mesh for the geometry, click on the Mesh 1 node in the Model Builder window. The Mesh window will open. In this window, select Fine from the list under Element size and click Build All. A total of 11,466 elements will be built, as displayed in the Messages window. Use Zoom Box to see the boundary elements adjacent to solid walls, as shown in Figure 4.94.

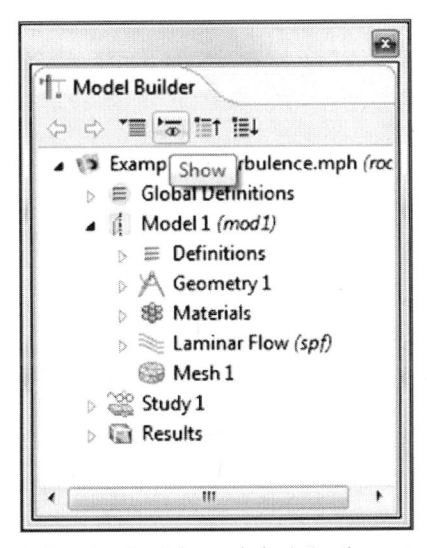

FIGURE 4.92 Model Builder window showing Advanced physics options setup.

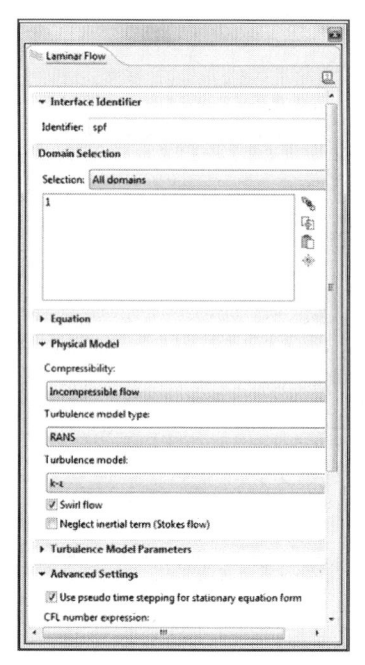

FIGURE 4.93 Laminar flow window showing k-ε turbulence model setup.

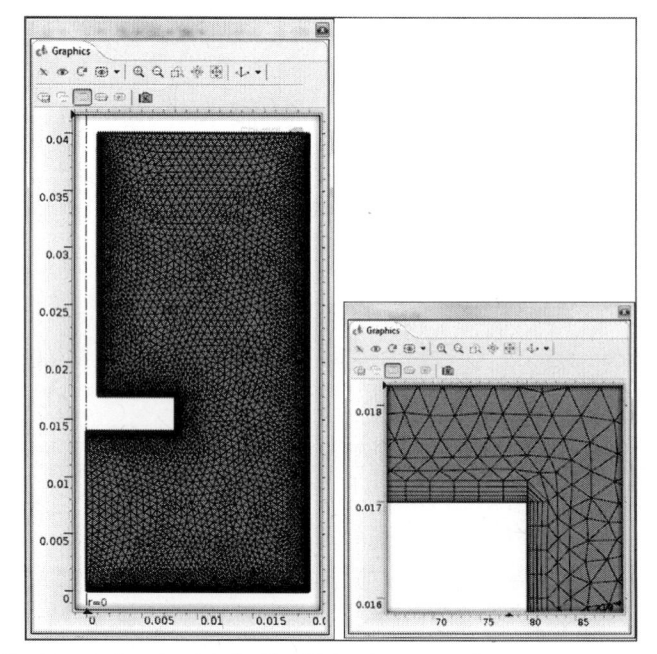

FIGURE 4.94 Graphics window showing details of finite elements mesh.

5. To set the values for omega, click on Step1: Stationary under the Study 1 node in the Model Builder. The Stationary window will open. In this window, expand the Study Extensions section and in the table enter the following data for omega:

Continuation parameter	Parameter value list
omega (Disk angular velocity)	range(100*pi,200*pi,500*pi)

Omega values are then set to start from 100π to 500π with the step of 200π.

6. To run the model, right-click on the Study 1 node and select Compute . Wait until the calculations are done. The default results should appear in the Graphics window, usually as 2D plots.

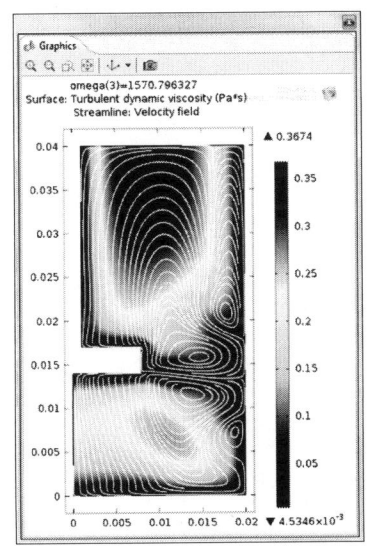

7. Reynolds stresses or turbulence viscous effects on the fluid are quantities of interest in a turbulent flow. To show this, right-click on Velocity (*spf*) and select Duplicate. The 2D Plot Group window will open. In this window, select 1570.796327 (the value for 500π) from the list for Parameter value (omega). Then expand the Velocity (*spf*) 1 node in the Model Builder window and click on Surface 1. In the Surface window, click the Replace Expression icon and select Turbulent Flow, $k - \varepsilon$ > Turbulent dynamic viscosity (spf.muT) from the list. Also in this window, expand the Coloring and Style section and check the Color legend box. Finally, click on Plot . The results, as shown in Figure 4.95, display

FIGURE 4.95 Results for turbulent dynamic viscosity and surface velocity for omega = 500π.

Ω (Rad/s)	μ (Pa.s)	μ_{Tmax} (Pa.s)	Re	μ_{Tmax}/μ
100π	0.001	0.0765	2.0106E4	76.5
300π	0.001	0.223	6.0319E4	223
500π	0.001	0.3674	1.005E5	367.4

TABLE 4.2 Turbulent dynamic viscosity and flow Reynolds number.

turbulent dynamic viscosity mapped with streamlines for disk angular velocity of 500π. The value of turbulence dynamic viscosity ranges from 0.0765 to 0.3674 Pa.s. Comparing this to the fluid dynamic viscosity 1e-3 shows the importance of turbulence viscous effects when flow becomes turbulent (see Table 4.2).

8. It is useful to calculate the Reynolds number for different values of disk angular velocity as well as the Cell Reynolds number. We defined the Reynolds number is calculated for fluid kinematic viscosity (91E-6) and r is 0.008 m. Table 4.2 summarizes these results.

Example 4.10: Flow in a U-shape Pipe with Square Cross-sectional Area: Laminar Flow

In this example, we model the flow in a pipe or duct with square cross-sectional area. Calculations for flow, mainly velocity and pressure, in pipes are needed in many engineering applications like ventilation and air conditioning ducts. Semi-empirical formulae are available and commonly used for these types of calculations. For noncircular pipes, usually engineers calculate the equivalent hydraulic diameter and use the circular pipe formulations. For reference, hydraulic diameter is defined as $D_h = \frac{4A}{P}$, where A is the area of pipe cross-section and P is the (wetted) perimeter. COMSOL 4.3 has a new module for pipe network calculations and modeling that assumes 1D flow in the pipes and is a useful tool for modeling pipe networks. However, sometimes detailed 3D flow modeling is needed for engineering and industrial applications.

Solution

1. Launch COMSOL and open a new file. Save the file as Example 4.10.
2. Select 3D from the list in the Model Wizard window and click Next ⇨. In the same window from the Add Physics list, expand the list for

Fluid Flow and select Laminar Flow (*spf*), Fluid Flow>Single-Phase Flow>Laminar Flow (*spf*). Click on the plus sign ➕ to add it to the Selected physics list. Click on Next ➡ and select Stationary from the list under Select Study Type. Click on the flag icon to Finish 🏁.

3. Now we draw the geometry of the pipe in COMSOL. You can also draw it using a CAD package and import it into this model. In the Geometry window, change the Length unit to mm by selecting it from the list. Right-click on the Geometry node in the Model Builder window and select Work Plane from the list. Click on the Plane Geometry node, which is created under Work Plane 1 (wp1) node in the Model Builder window. See Figure 4.96.

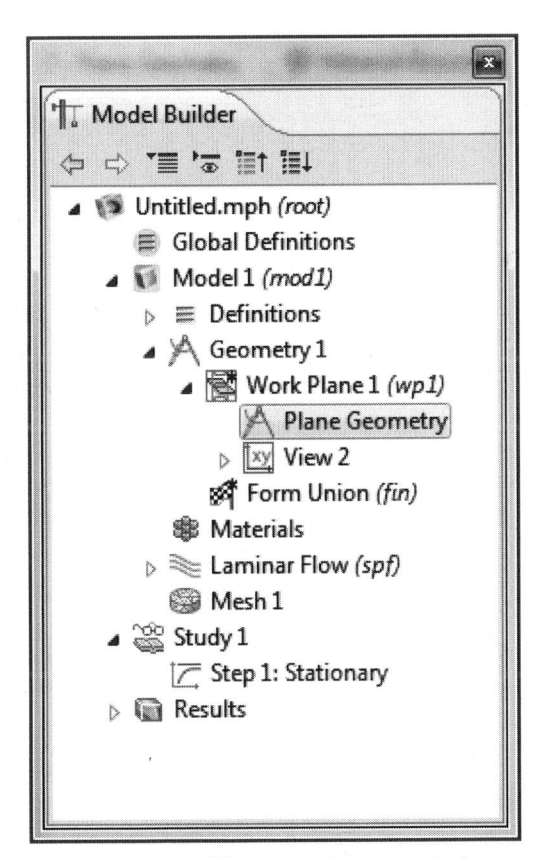

FIGURE 4.96 Model Builder window showing work plane setup.

4. In the Graphics window, clicking on the square icon in the main toolbar and draw a square. In the Model Builder, expand the list under Plane Geometry and click on the Square 1 (*sq1*) node. The Square window will open. In this window, change the data as shown in Figure 4.97, and then click the Build Selected icon . The square with the given size and corner coordinates will appear in the Graphics window. In this window, click the Zoom Extents icon to adjust.

5. Now we extrude this square to build the pipe geometry. Right-click on the Work Plane 1 (*wp1*) node in the Model Builder Window and select Extrude from the list. The Extrude window will open. Locate Distance from the Plane and type 200 in the space under Distance. This will set the length of the pipe. Click the Build Selected icon . Click the Zoom Extents icon to adjust. See Figure 4.98.

6. Create another work plane on the top surface of the pipe. Right-click on the Geometry node in the Model Builder window and select Work Plane from the list. Click on the Work Plane 2(*wp2*) node in the Model Builder window. In the Work Plane window, change the Plane type and Origin as shown in Figure 4.99, and then click the Build Selected icon .

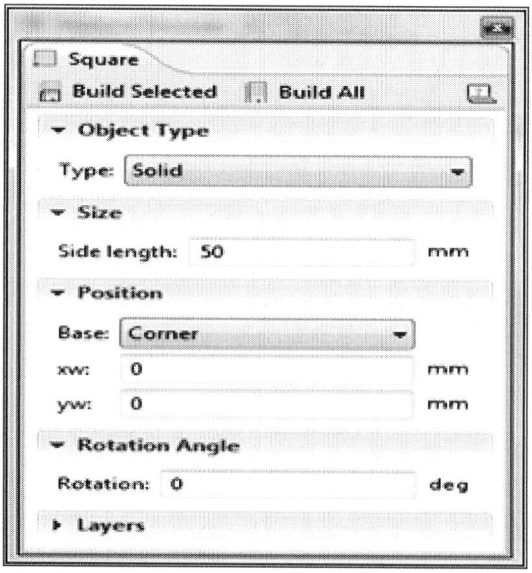

FIGURE 4.97 Square geometry set up.

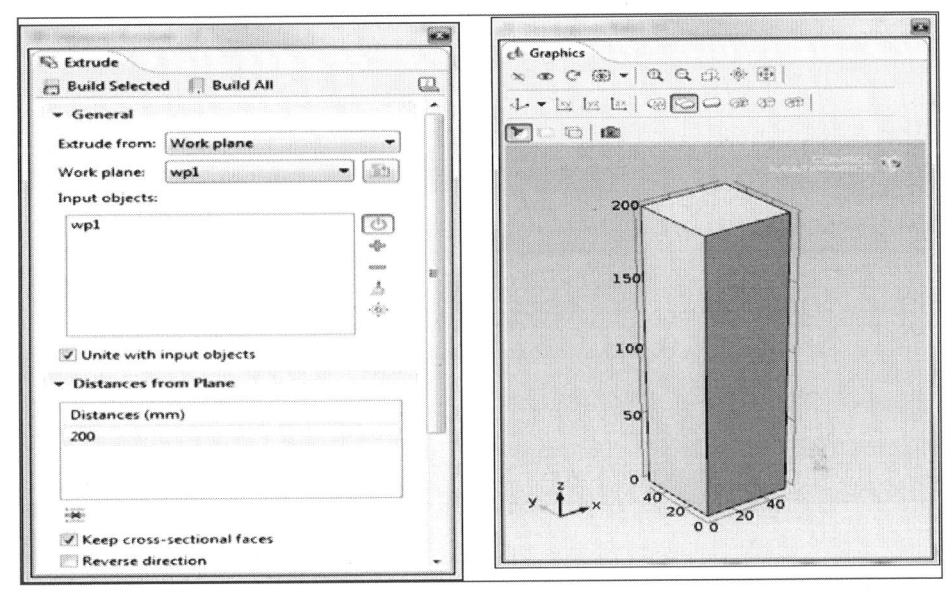

FIGURE 4.98 Windows showing Extrude set up and resulting geometry.

FIGURE 4.99 Work Plane setup.

122 • COMSOL FOR ENGINEERS

7. Click on the Plane Geometry node, which is created under Work Plane 1 (*wp1*) node in the Model Builder window. Draw a square in the Graphics window by following steps similar to those described in Step 4 above.

8. To build the bend, right-click on Work Plane 2 (*wp2*) and select Revolve from the list. The Revolve window will open. In this window, locate the Revolution Angles and Revolution Axis sections, input the data as shown in Figure 4.100, and then click the Build Selected icon .

9. To build the last part of the pipe, right-click on the Geometry 1 node in the Model Builder window and select Transforms>Copy. The Copy window will open. In this window, add the straight section of the pipe (which was built previously) to the Input objects list by clicking on it in the Graphics window and then right-click. In the Displacement, enter 70 for x and click on the Build Selected icon . See Figure 4.101.

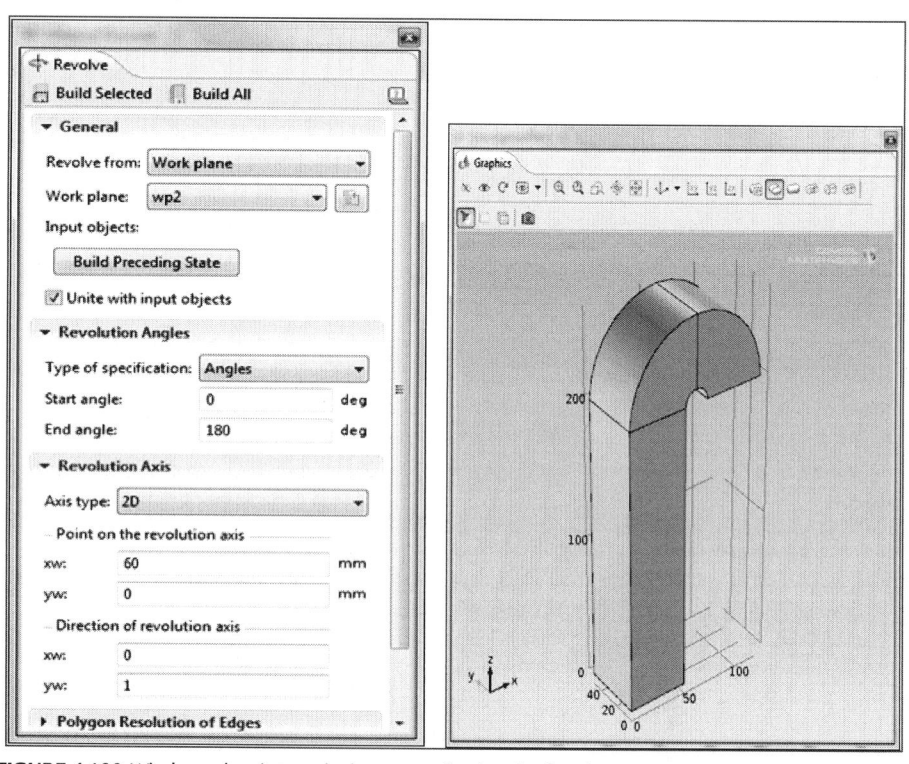

FIGURE 4.100 Windows showing work plane setup for the pipe bend.

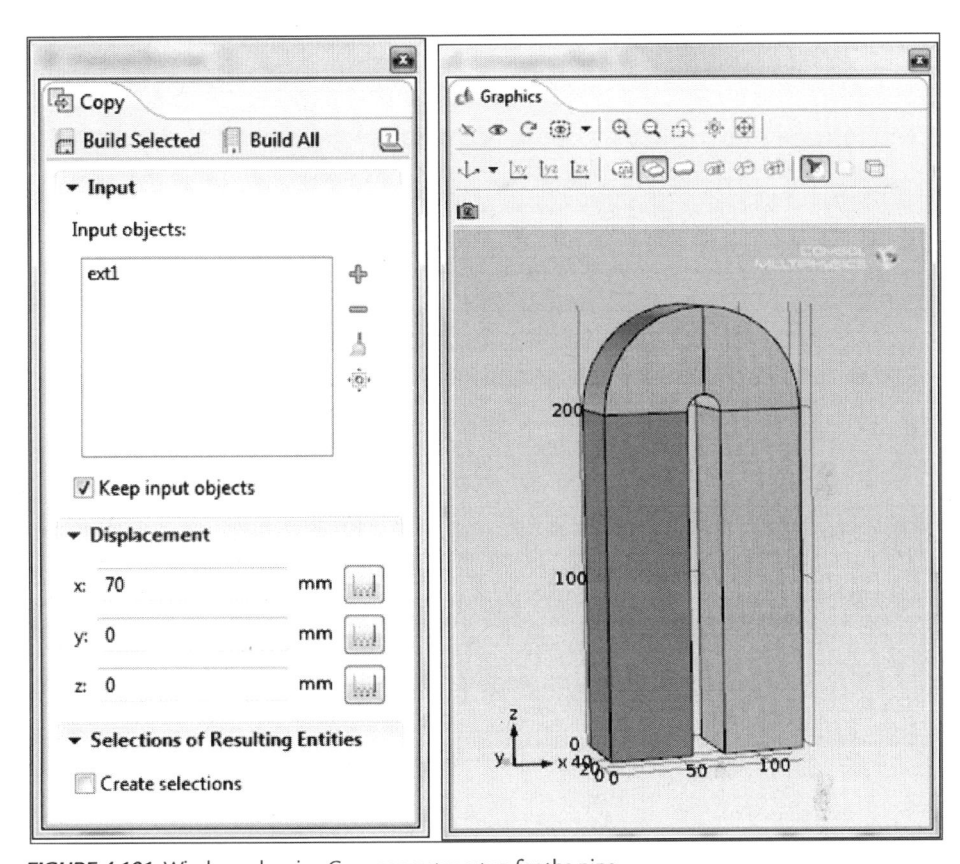

FIGURE 4.101 Windows showing Copy geometry setup for the pipe.

10. We now add materials to the pipe domain. Right-click on the Materials node in the Model Builder window and select Open Material Browser from the list. In the corresponding window, locate Liquids and Gases>Gases>Air. Right-click on Air and select Add Material to Model.

11. For the air flow we define a parameter as **Vmax** to be the maximum velocity at the entrance to the pipe. Right-click on the Global Definitions node in the Model Builder and select Parameters. In the corresponding window, type the data Name: **Vmax**, Expression: 1[m/s], and Description: **Inlet max. velocity.**

12. We now define the boundary conditions at the pipe inlet and outlet. The no-slip boundary condition is set by default for other pipe walls.

   12.1 For defining the boundary condition at the inlet, right-click on the Laminar Flow (*spf*) node in the Model Builder and select Inlet from the list. In the Inlet window, add one of the end surfaces (3) of the pipe to the Selection, click on the surface in the Graphics window, and right-click again. It is optional but useful to expand the Equation section, which shows the governing equations for the type of boundary conditions selected. Here we select Laminar inflow from the list in the Boundary Condition. Select Average velocity and enter **0.5*Vmax**. For the Entrance Length, enter 20 for L_{entr}. Also, check the box for Constrain outer edges to zero. This will force the laminar velocity profile to go to zero at the edges of the pipe inlet. A technical point should be mentioned here. COMSOL uses the value of L_{entr} to calculate a fully developed velocity outside the domain of the model, as if you have added an extension to the entrance. This is a very useful feature in COMSOL. The value of L_{entr} should be larger than $0.06ReD$ (see COMSOL Manual), where Re is the Reynolds number and D is the pipe hydraulic diameter. For our model, we have $(0.06)(10^4/3)(0.05) = 10$ m, when Vmax = 1 m/s, air kinematic viscosity is 1.5×10^{-5}, and D = 0.05 m. For fully developed flow it can be shown that maximum velocity is 1.5 times the average velocity. See Figure 4.102.

   12.2 For defining outlet condition, right-click on the Laminar Flow (*spf*) node and select Outlet. In the corresponding window, add the outlet surface (17) of the pipe to the Selection list.

13. Build a mesh by clicking on the Mesh node in the Model Builder window, and then select Coarser for the Element size in the Mesh window. Make sure that Sequence type is set to Physics-controlled mesh. Click the Build All icon. A total of 32,557 finite elements will be built. Results are shown in Figure 4.103.

   Notice the hybrid mesh types, which include boundary-layer hexagonal and tetragonal elements for the internal flow region.

14. Next, we add a parametric study based on **Vmax**. Right-click on the Study1 node in the Model Builder window and select Parameter Sweep. In the corresponding window, locate the Study settings and click on the plus sign ✚ to add the existing parameter **Vmax** to the

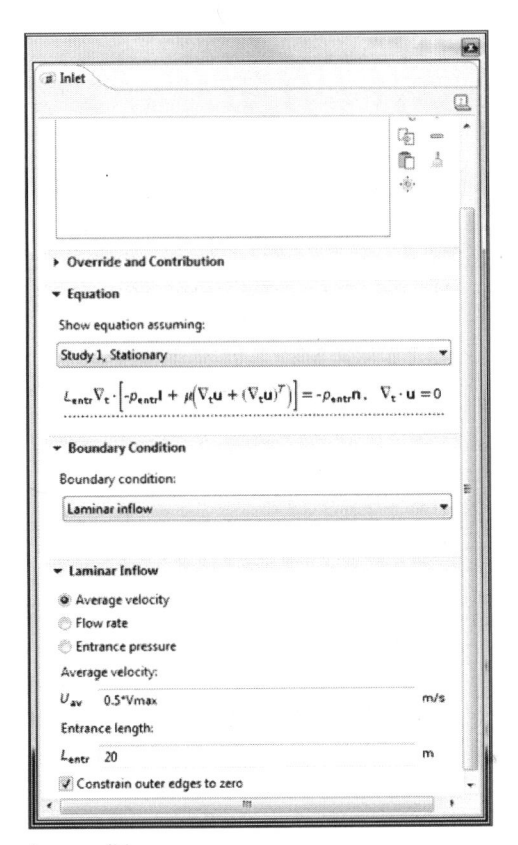

FIGURE 4.102 Inlet boundary condition setup.

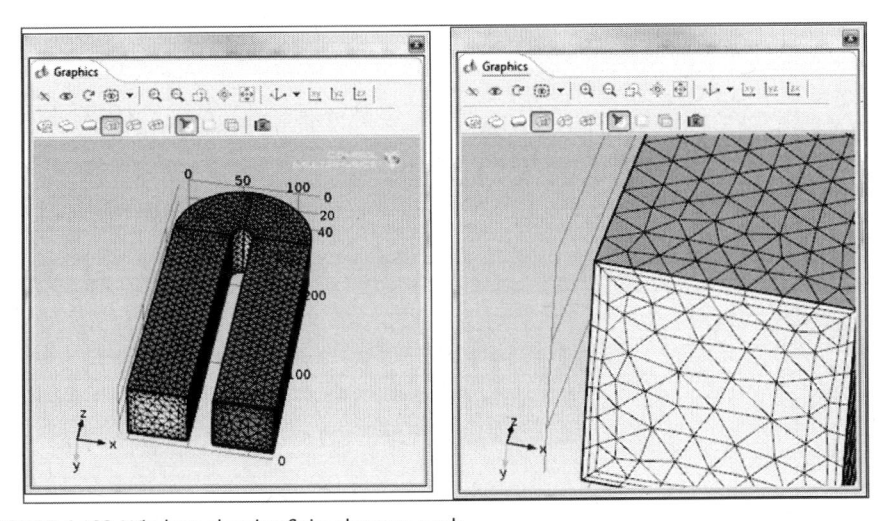

FIGURE 4.103 Windows showing finite elements mesh.

list in the table. Enter 0.5, 1, 1.5 for the Parameter value list, as shown in Figure 4.104. Notice that these values should satisfy the criteria $L_{entr} > 0.06ReD$.

15. Run the model, right-click on the Study 1 node in the Model Builder window, and select Compute ≡. Alternatively, the equal icon ≡ in the main toolbar can be clicked. Wait until computations are finished.

16. To manipulate the default results, expand the Velocity (*spf*) node and click on Slice 1. In the corresponding window, locate Plane Data and select xy-planes from the list. Click the Plot icon. See Figure 4.105. The results will show the air velocity magnitudes for the Vmax = 1.5 m/s. Notice the effect of the bend on the flow, especially through the downstream section of the pipe, as shown in Figure 4.106.

To study the results, it would be useful to use 1 plane in the Interactive mode. Enter 1 for Planes and check the box for Interactive. By sliding the slider nub, the plane will move through the pipe domain while showing the modeling results for velocity magnitude.

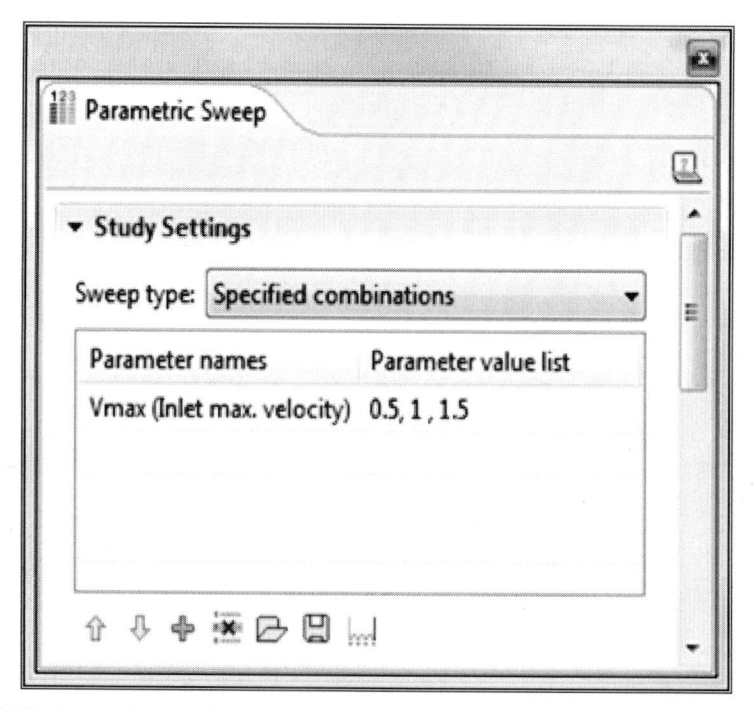

FIGURE 4.104 Parametric values for inlet velocity.

FIGURE 4.105 Plot setting for flow velocity magnitude inside the pipe.

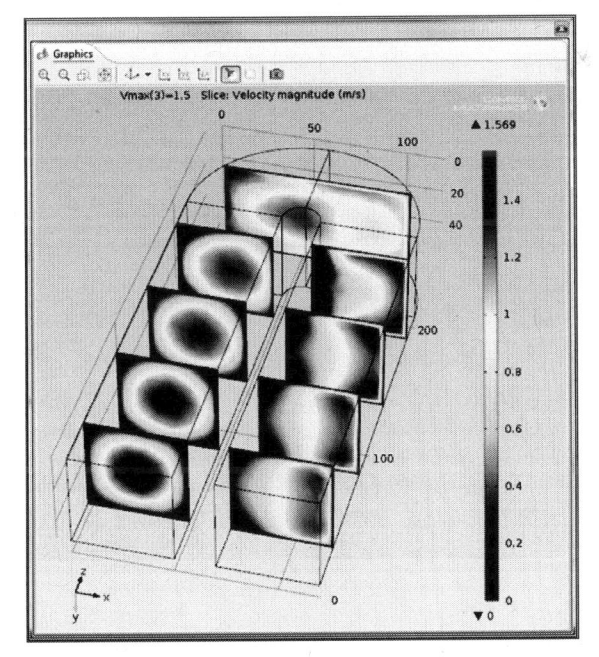

FIGURE 4.106 Flow velocity along the pipe cross-sections.

17. To show the results through the bend we create a Selection out of the domain and the Results, which is limited to the volume inside the pipe bend. Expand the Data Sets node in the Model Builder window, right-click on the Solution 1, and select Add Selection. Click on the newly created Selection node and in the corresponding window select Domain from the list for Geometric entity level. Click on the domain that represents the bend (i.e., domain 2) in the Graphics window and then right-click to add it to the Selection list. See Figure 4.107.

Now click on the Slice 1 under Velocity (*spf*) node and in the corresponding window under Plane Data change the Plane to yz-planes. The velocity magnitude will be displayed in the bend domain only, as shown in Figure 4.108.

18. To show the results for other values of Vmax, click on Velocity (*spf*) node in the Model Builder. In the corresponding window, locate the Data section and select the desired Parameter value (*Vmax*). These values are the same as those set for these parameters.

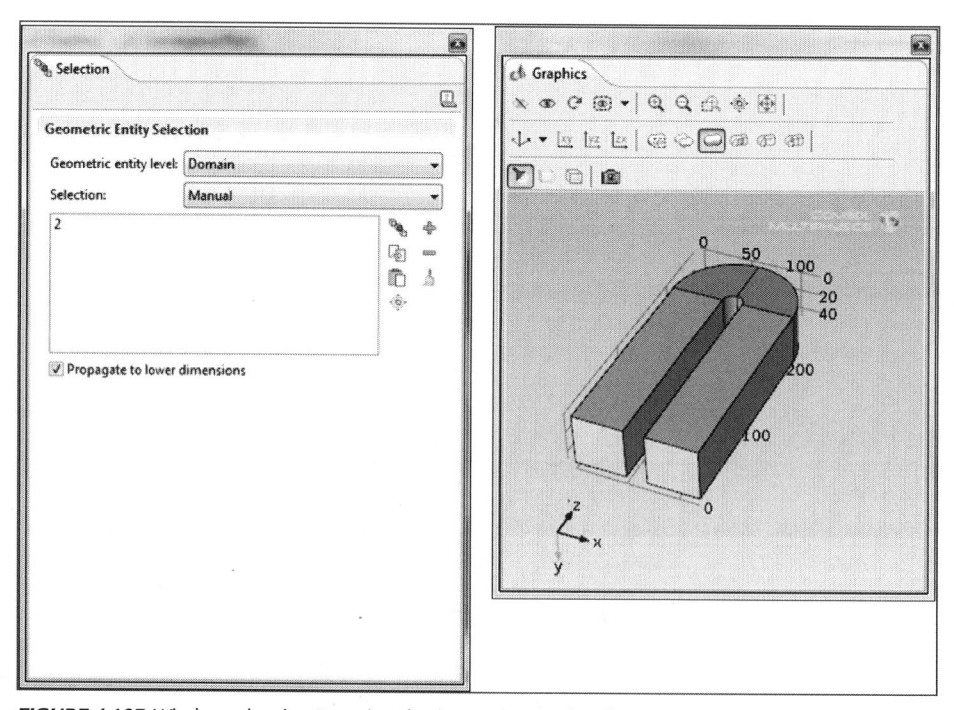

FIGURE 4.107 Windows showing Domain selection as the pipe bend.

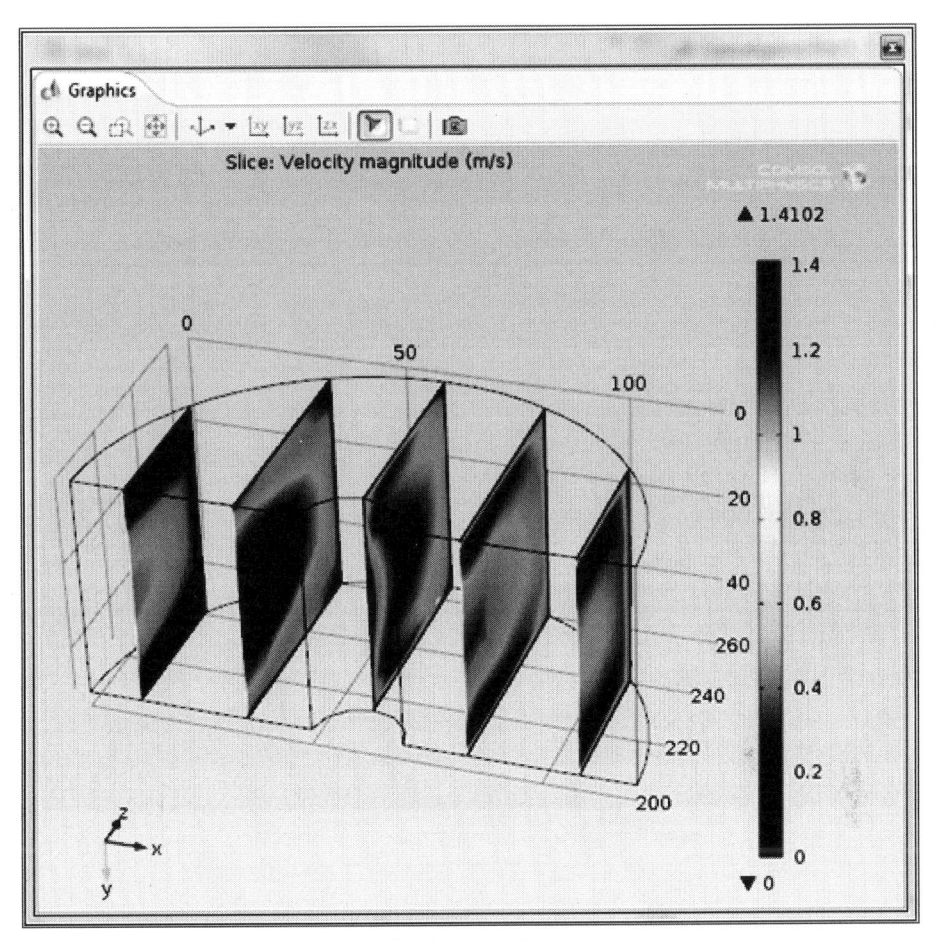

FIGURE 4.108 Flow velocity along the pipe bend cross-sections.

Example 4.11: Double-driven Cavity Flow: Moving Boundary Conditions

Driven cavity is a classic example of flow in a square shape geometry (2D) in which one of the square's side is a moving wall with a constant speed. Double-driven cavity is a version of this problem with two partially merged domains or squares and two walls moving in opposite directions (see Reference 4.6). In this example, we use COMSOL to find the solution for double-driven cavity. Cavity flow situations can happen in nature, such as in estuaries, and in industrial flows, such as mixing tanks.

Solution

1. Launch COMSOL and open a New file. Save the file as Example 4.11.

2. Select 2D in the Model Wizard window and click Next ➪.

3. In the Add Physics section, select Fluid Flow>Single-Phase Flow>Laminar Flow (*spf*) and click on ✚ to add it to the Selected Physics list. Click ➪ to move to the next step.

4. Select Stationary from the list under Study and click the flag icon 🏁 to Finish.

5. We will draw the geometry of the double-lid driven cavity. In the Geometry window, change the Length unit to **cm**. Next, we define parameters for size of the length of the cavity and speed of its lids. Right-click on the Global Definitions and select Parameters. In the Parameters window, enter the data as shown in Figure 4.109. The Reynolds number is defined as $Re = \dfrac{V_{wall} \cdot l}{v}$, where V_{wall} is the lid absolute speed, l is the cavity size, and v is the kinematic viscosity of the fluid inside the cavity.

6. Right-click on the Geometry node in the Model Builder window and Select Square from the list. In the Square window, enter L for the Side length under Size section, and then click the Build Selected icon 🖫. A square with size L will appear in the Graphics window. Draw another square by right-clicking again on the Geometry node in the Model Builder and selecting Square. In the corresponding window, enter L for the Side length and 0.4*L for both x and y under Position. Click 🖫. To adjust the window, click on ✥ located in the toolbar of the Graphics window. Results are shown in Figure 4.110.

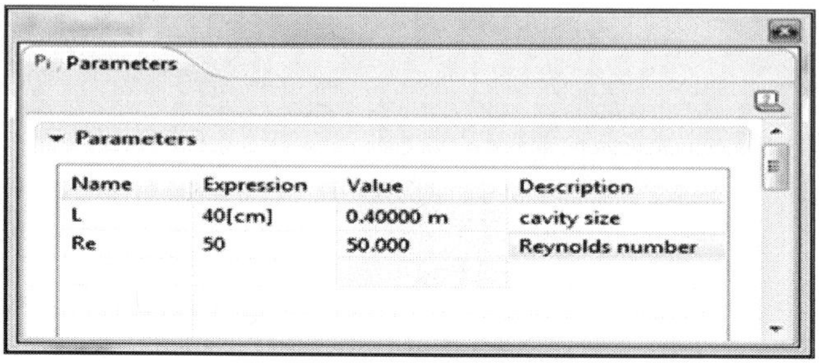

Name	Expression	Value	Description
L	40[cm]	0.40000 m	cavity size
Re	50	50.000	Reynolds number

FIGURE 4.109 Parameters setting.

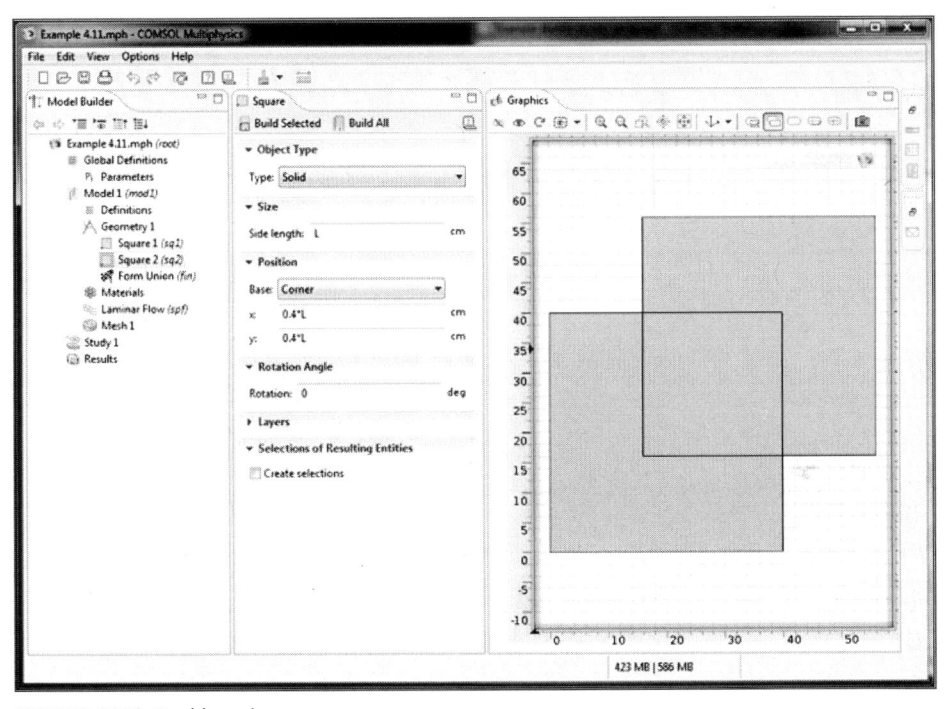

FIGURE 4.110 Double-cavity geometry set up.

7. To define a material for the fluid, right-click on Materials and select Open Material Browser from the list. In the corresponding window, click and select Liquid and Gases>Liquids>Water. Click on the ✚ and Add Material to Model. Water will be assigned to the whole domain of cavity space automatically. Notice the check marks that appear in the Material window under Material contents for Dynamic Viscosity and Density properties. Also notice that these quantities will be calculated for a given temperature from the database. See Figure 4.111.

8. We define the boundary conditions now. Right-click on the Laminar Flow (*spf*) node in the Model Builder window and select Wall from the list. A Wall 2 node will appear in the model tree. Right-click Wall 2, select Rename from the list, and change the name to Upper_Lid. Repeat this last operation to create another wall and rename it to Lower_Lid. Click on the Upper_Lid node in the Model Builder, and then click on the upper edge of the cavity in the Graphics window. Add this edge to the Selection by clicking on the ✚. Boundary number 8

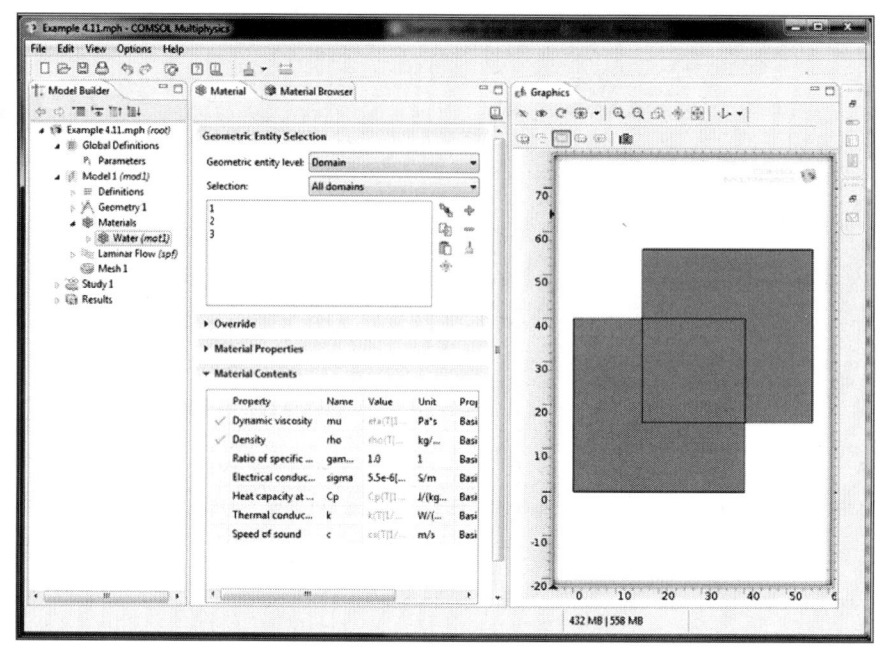

FIGURE 4.111 Adding water properties to the model.

should appear in the Selection list in the Wall window. In the same window, select Moving wall from the list under Boundary condition and enter $(Re \times spf.mu)/(L \times spf.rho)$ for x. This expression defines the speed of the upper lid, as shown in Figure 4.112.

Similarly, for Lower_Lid (i.e., boundary number 2) define the speed as $(Re \times spf.mu)/(L \times spf.rho)$. The reaming walls are defined as no-slip by default; no operations are needed to define them. We should define a point for pressure reference since the flow inside the cavity is a closed one. Right-click on Laminar Flow (*spf*) and select Points>Pressure Point Constraint. In the corresponding window, click on point 7 in the cavity geometry and add it to the list of Selections by right-clicking. See Figure 4.113.

9. For complex flow models like this example, it is recommended to define a few other features to enhance the convergence of the solution. To show the advanced physics options, click on the Show icon in the toolbar of the Model Builder window and select Advanced Physics Options. Then click on the Laminar Flow (*spf*) node and in the

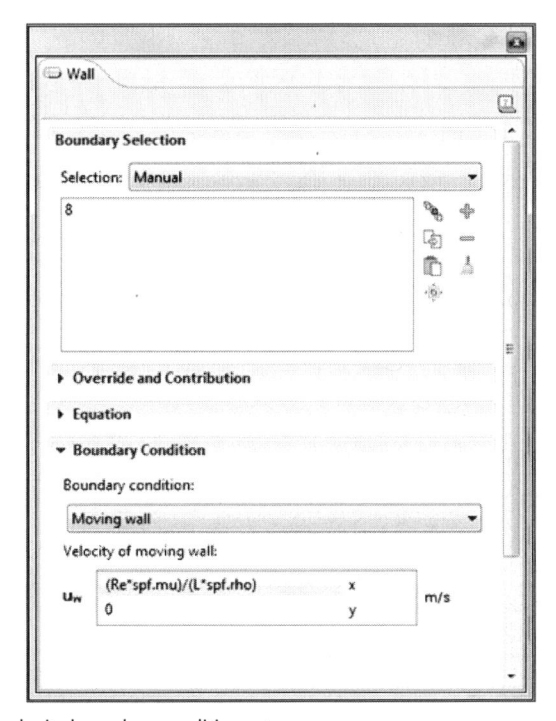

FIGURE 4.112 Wall velocity boundary condition setup.

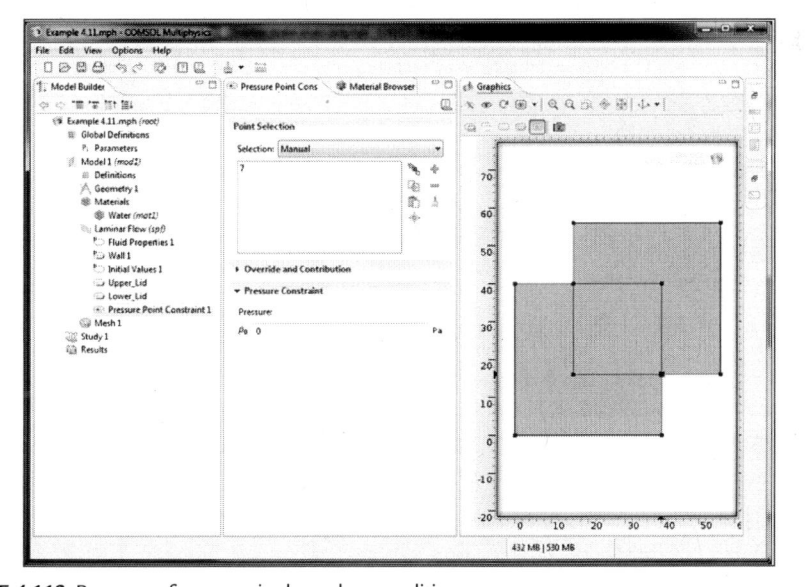

FIGURE 4.113 Pressure reference point boundary condition setup.

corresponding window locate Advanced Settings section and expand it. Check the box for *Use pseudo time stepping for stationary equation form*. Also in this window, select Incompressible flow from the list under Compressibility in the Physical Model section. See Figure 4.114.

10. To build a mesh, click on the Mesh in the Model Builder window. In the Mesh window, accept all default values and click on the Build All icon . A mesh consisting of 1252 elements will be built inside the flow domain. To obtain the mesh statistics and quality, right-click on Mesh 1 and select Statistics. See Figure 4.115.

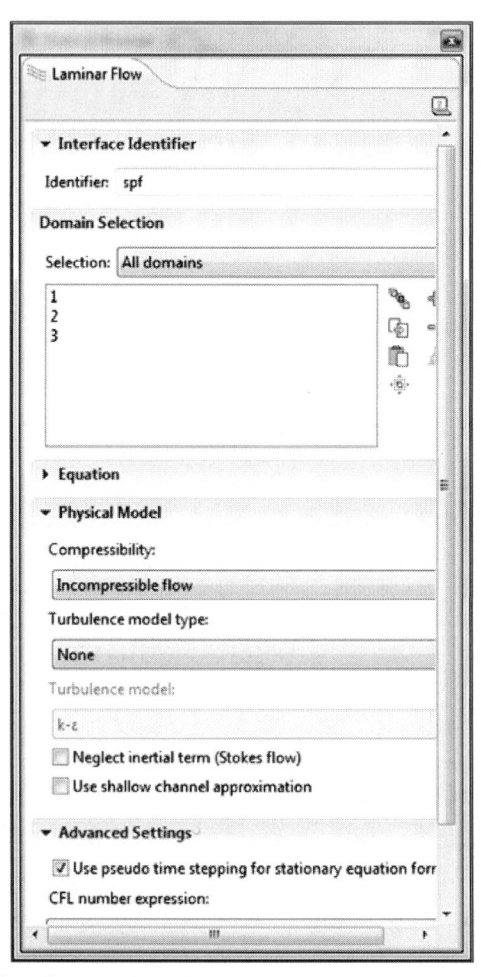

FIGURE 4.114 Laminar flow physics and pseudo time stepping setup.

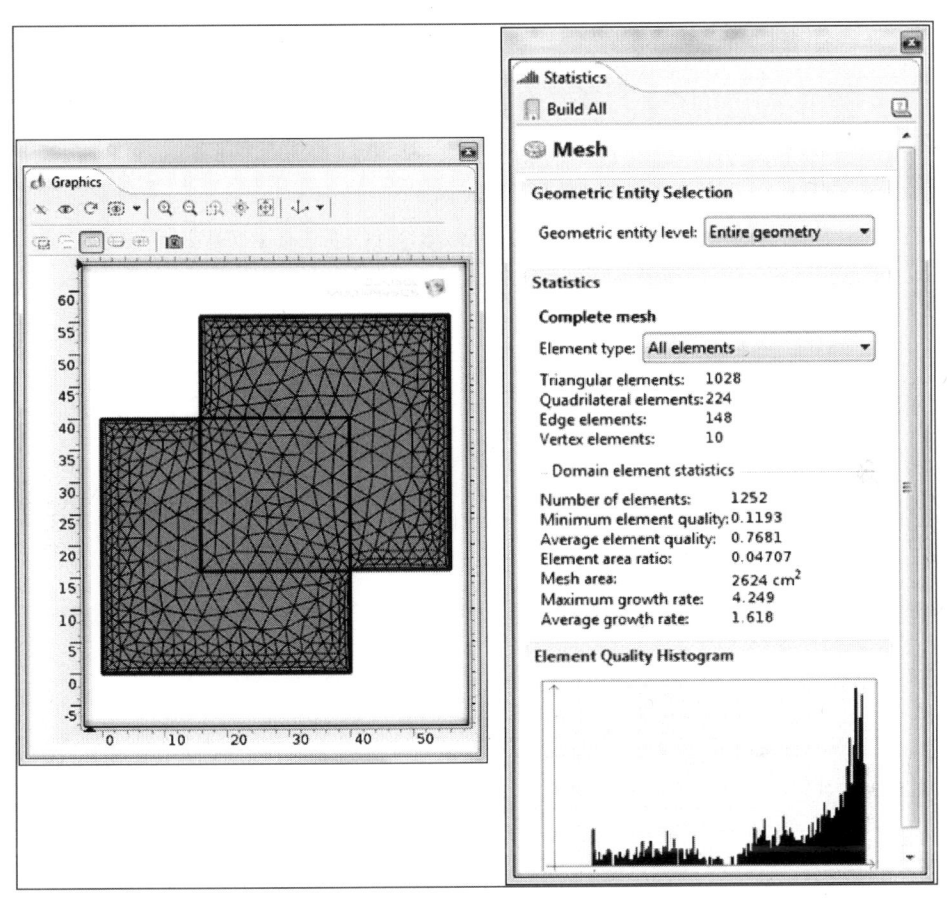

FIGURE 4.115 Finite elements mesh and statistics.

11. To run the model, right-click on the Study 1 node in the Model Builder window and select Compute ≡. Wait for the computations to finish. The default results will appear in the Graphics window, as shown in Figure 4.116.

12. To distinguish the solution, we rename it. Expand Data Sets, located under Results node, and rename Solution 1 to **SolutionRe50**, to indicate that these are the solutions for Re equal to 50. Also, by default the inner boundaries are shown in the surface plots. To eliminate these inner boundaries, right-click on the SolutionRe50 node and select Add Selection. The Selection window will open. In this window, select Boundary from the list for Geometric entity level. From the Selection

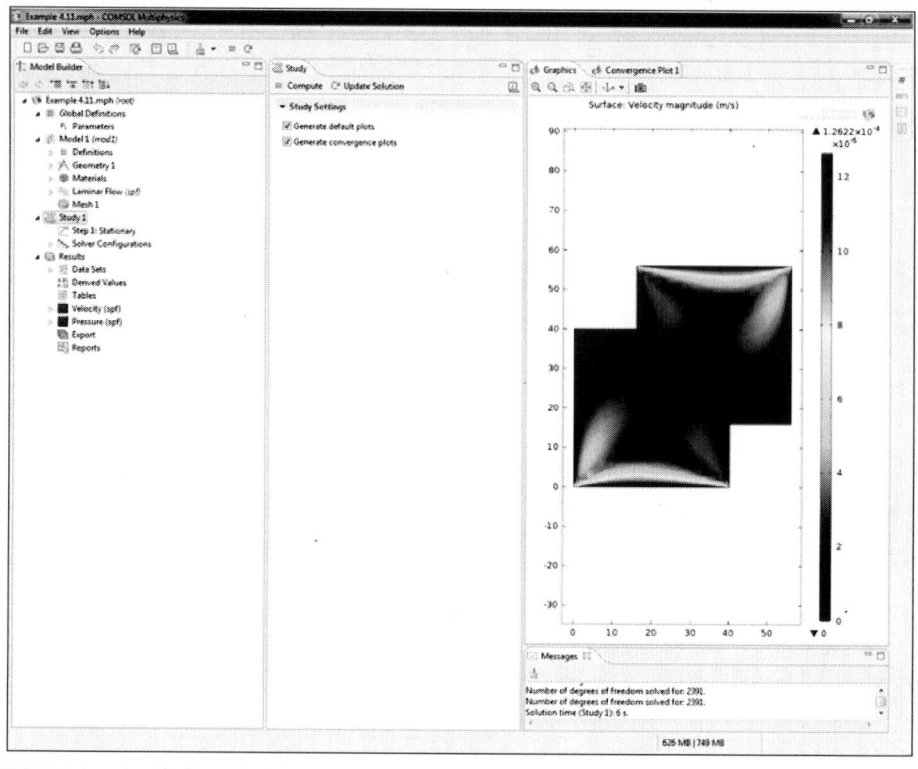

FIGURE 4.116 Default results for flow velocity in double cavity.

list, choose All boundaries. Click on the four inner boundaries, one by one, and then click on the minus icon to eliminate them from the list. Click on the Velocity (*spf*) to see the results, which now do not show the inner boundaries. See Figure 4.117.

13. To analyze the effect of mesh resolution on the results, we create another mesh with higher resolution. Rename the Mesh 1 to Mesh-Normal. Right-click on the MeshNormal node in the Model Builder window and select Duplicate from the list. A new Mesh node will be created in the model tree; rename it to MeshFiner. In the corresponding Mesh window, select Finer for the Element size and click the Build All icon . A total of 5810 finite elements will be created. Now we have two mesh sets, which can be shown in the Graphics window when clicking on MeshNormal or MeshFiner nodes. See Figure 4.118.

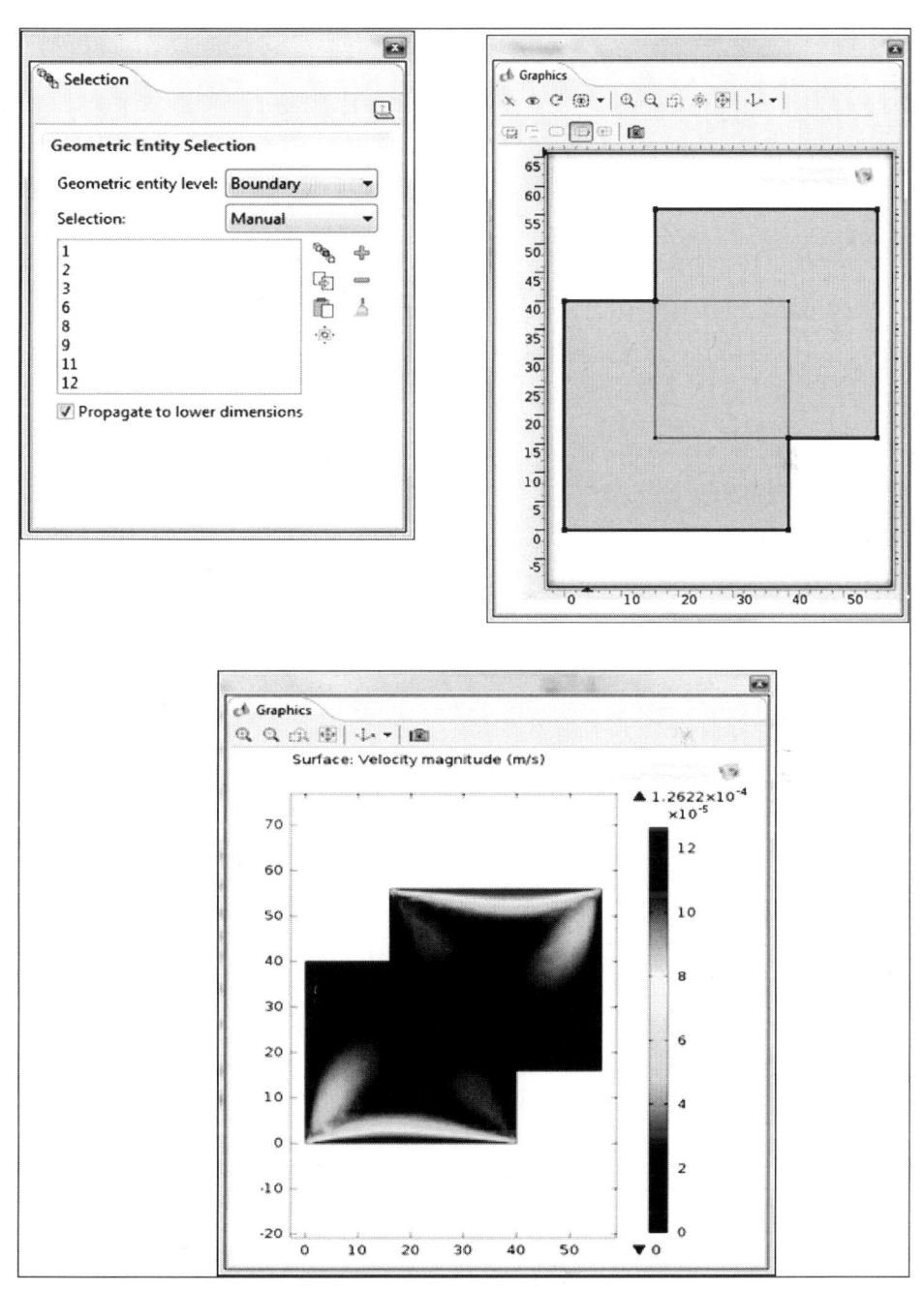

FIGURE 4.117 Windows showing selection of domains and flow velocity results.

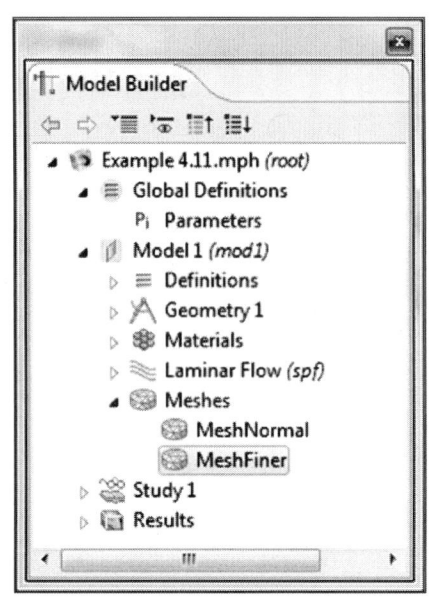

FIGURE 4.118 Model Builder window showing mesh options setup.

14. Expand Results and right-click Data Sets>Solution to add a new Solution 2 as data subset. Rename Solution 1 as SolutionMeshNormal and Solution 2 to SolutionMeshFiner. To run the model using the finer mesh, click on the Step 1: Stationary node, located in Study 1 in the model Builder Window. In the corresponding window, expand the Mesh Selection section and select MeshFiner from the list under Mesh. Right-click on the Study 1 node and select Compute. Wait for computations to finish. See Figure 4.119.

15. Now we have solutions with normal mesh stored in SolutionMeshNormal and those for finer mesh in SolutionMeshFiner. We create another Data set that stores the difference between these two solutions. Right-click on Data Sets and select Join from the list. Click on the newly created node Join 1. In the corresponding window, select SolutionMeshNormal from the list for Data 1, select SolutionMeshFiner for Data 2, and then select Difference for Combination Method. See Figure 4.120.

16. We create a new plot group to graph the results stored in Join 1. Right-click on the Results node and select 2D Plot Group. A new

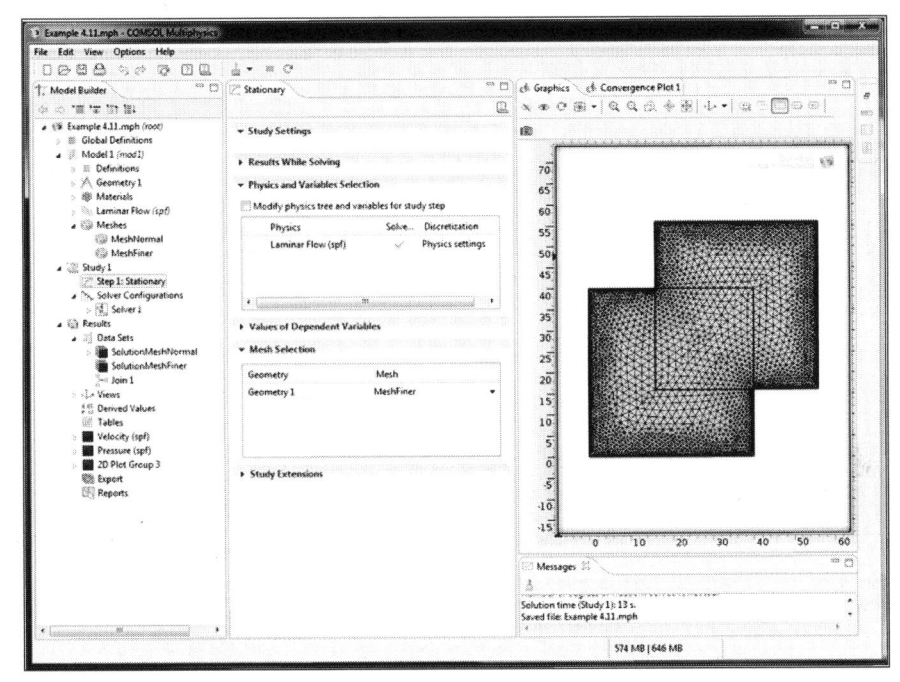

FIGURE 4.119 Finite elements mesh for MeshFiner option.

FIGURE 4.120 Combination option Difference set up for two solutions.

plot group will appear as 2D Plot Group 3; right-click on it and select Surface. The Surface window will appear; in this window under Data, select Join 1 for Data set. Also expand Coloring and Style and change the options as shown in Figure 4.121. Click on Plot to plot the results. The difference between the two solutions is very small and negligible.

17. Next, we run the model for several values of Reynolds number. We use MeshNormal for these calculations. Disable SolutionMeshFiner and Join 1 by right-clicking on them and selecting Disable from the list. Click on the Step 1: Stationary node in the Model Builder window. Locate Mesh Selection and select MeshNormal under Mesh list. Expand Study Extensions and check the box for Continuation. Click on the plus icon ➕ and from the list of parameters under Continuation parameter, select Re (Reynolds number). For its values enter 50, 100, 400, 1000 in the space under the Parameter value list, as shown in Figure 4.122.

18. Right-click on the Velocity node in the Model Builder and select Contour. In the Contour window, type **u** in the space available under Expression. To draw the contours for a different Re, click on Velocity (*spf*) and select a value for the Parameter value (Re). The results for Re = 50, 100, 400, 1000 are shown in Figure 4.123.

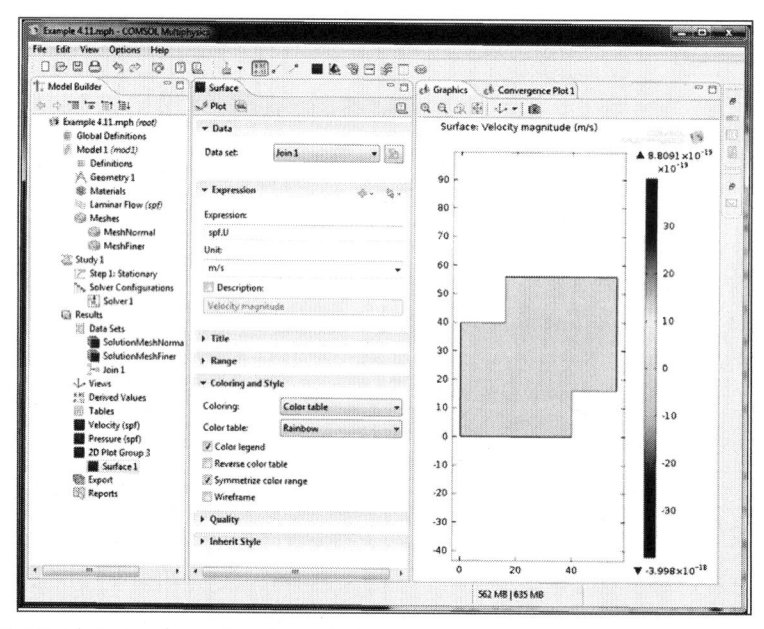

FIGURE 4.121 Join1 setup for resulting flow velocity.

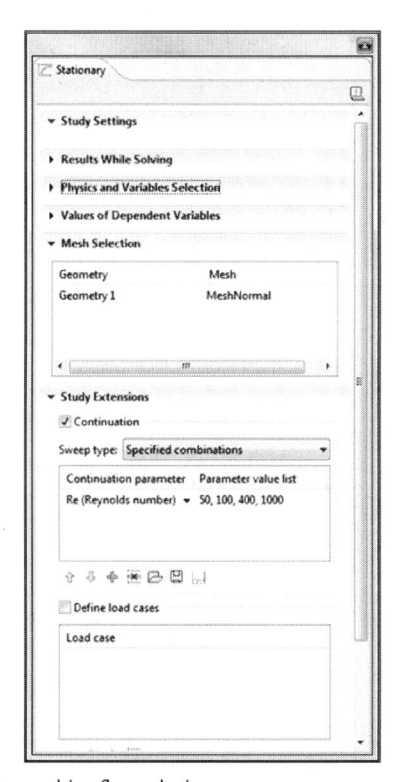

FIGURE 4.122 Join1 setting for resulting flow velocity.

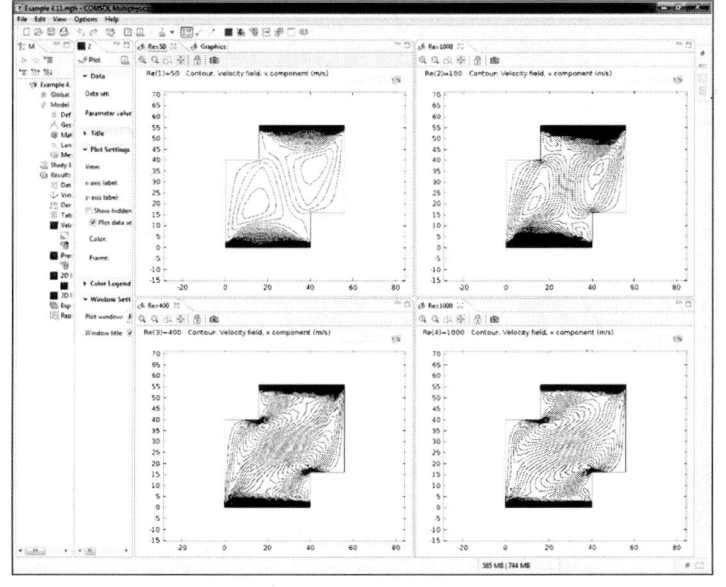

FIGURE 4.123 Windows showing results for flow x-component of velocity for several Reynolds numbers.

Example 4.12: Water Hammer Model: Transient Flow Analysis

Water hammer is a phenomenon that occurs as a result of a sudden pressure change or pressure pulse, usually in pipelines that conduct water, but it can also occur in steam or multiphase fluids conduits. For this example, we consider water as the fluid moving in pipes. The pressure pulse could be the result of sudden closure of a valve or some other changes in the fluids that result in a pressure wave propagation. The pressure pulse creates a pressure wave that propagates through water and the pipe material. For many applications, water can be considered an incompressible fluid; for water hammer, water compressibility should be considered for a more exact solution to the problem. Deformation of the pipe material affects the pressure wave propagation and should be considered as well. Water hammer governing equations can be derived from Navier-Stokes equations for compressible fluids (see Reference 4.7). The speed of pressure wave propagation is much larger than average water velocity and a 1D mathematical model is usually used for modeling:

$$\frac{g}{c^2}\frac{\partial H}{\partial t} + \frac{\partial V}{\partial x} = 0$$

$$\frac{\partial V}{\partial t} + g\frac{\partial H}{\partial x} + \frac{\tau_w \pi D}{\rho A} = 0$$

where H is piezometric head, c pressure wave speed, V average fluid velocity, τ_w wall shear stress, g gravitational acceleration, A pipe cross-sectional area, and D diameter. In early cycles of water hammer and for many cases in practice the wall shear stress can be neglected, which simplifies the above equations, and pressure wave speed c is given by the following equation:

$$c = \left(\frac{d\rho}{dP} + \frac{\rho}{A}\frac{dA}{dP}\right)^{-0.5}$$

The first term in the bracket is the square of the inverse of speed of sound in water $c_0 = \sqrt{dP/d\rho}$. This term represents the compressibility of the liquid water. The second term in the bracket represents the effect of pipe material flexibility. Juokowsky's relation (see Reference 4.7) for water hammer could be derived from the 1D mathematical model equations as:

$$\Delta P = \pm \rho c V$$

where ΔP is the change in pressure. COMSOL 4.3 has a water hammer module that employs these equations. In this example, we use this module to model and analyze the results of a water hammer in a pipe network as shown in Figure 4.124, which has two measurement points designated as B-gauge and E-gauge.

Solution

1. Launch COMSOL and open a New file. Save the file as Example 4.12.
2. Check the 3D button and click Next ⇨. From the Add Physics list, select Fluid Flow>Single-Phase Flow>Water Hammer (*whtd*) and click the plus sign ✚ to add this to the Selected physics list. Click Next ⇨.
3. From the Selected Type list, select Time Dependent and click the Finish icon 🏁.
4. To define the parameters, right-click on the Global Definitions and select Parameters. In the Parameter window, enter the following data, as shown in Figure 4.125.

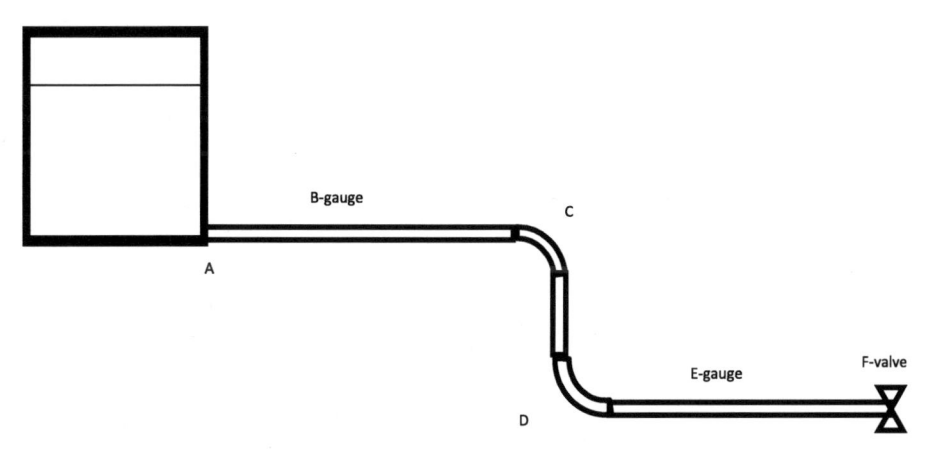

FIGURE 4.124 Geometry and components for a water network and storage.

Parameters

Name	Expression	Value	Description
Lac	10[m]	10.000 m	Pipe length
Lcd	3[m]	3.0000 m	Pipe length
Ldf	10[m]	10.000 m	Pipe length
GaugeB	5[m]	5.0000 m	Measurement point
GaugeE	12[m]	12.000 m	Measurement point
R	8[in]	0.20320 m	Pipe inner radius
w	8[mm]	0.0080000 m	Pipe wall thickness
E	210[GPa]	2.1000E11 Pa	Pipe Youngs modulus
Q0	0.5[m^3/s]	0.50000 m³/s	Initial flow rate
u0	Q0/(R^2*pi)	3.8545 m/s	Initial velocity
p0	1[atm]	1.0133E5 Pa	Initial pressure
L	10[m]	10.000 m	max (Lac,Lcd,Ldf)
N	400	400.00	Number of elements/m...
dt	0.2*L/N/1200[m/s]	4.1667E-6 s	Time step
dx	L/N	0.025000 m	

FIGURE 4.125 Parameters for water hammer quantities and their values.

The Expression for parameter time step **dt** should be explained here. Since the pressure surge is a sudden jump in pressure at a point, the mesh size and the time step should be carefully selected to have a stable and converged solution for the model. The pressure wave travels with velocity c and it takes $dt = dx/c$ seconds to travel through a distance dx, which is the mesh size. For a given typical pipe length L with number of elements N, we have $dx = L/N$. Then dt is equal to L/cN. However, we would like to have a much smaller time step so the wave can be captured within a mesh size, say:

$$dt = \frac{0.2L}{cN}$$

Speed of wave should be estimated, which is equal to 1200 m/s, here.

5. To draw the pipe network laying in the x-y plane, right-click on the Geometry 1 node in the Model Builder window and select More Primitives>Polygon. In the Polygon window, enter the following

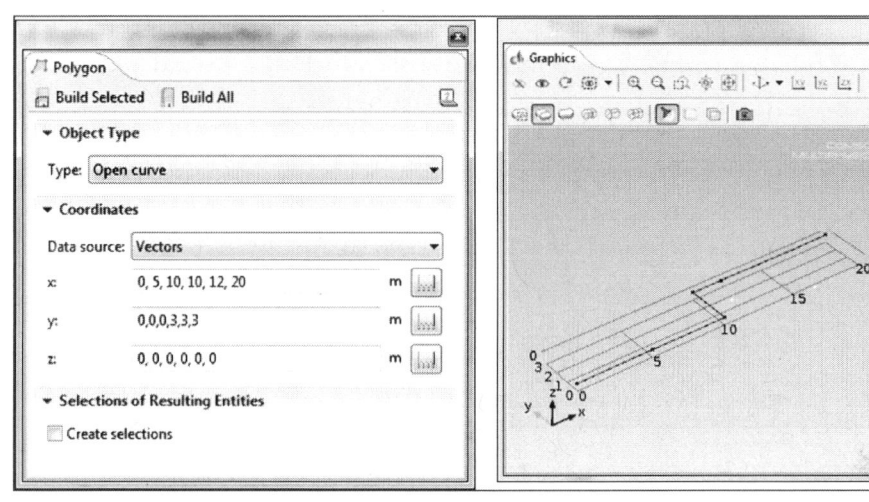

FIGURE 4.126 Values for building the pipeline geometry and results.

data in the Coordinates section and click Build All . The pipeline geometry will appear in the Graphics window. Nodes at x = 5 and 12 are measurement/gauge points. Results are shown in Figure 4.126.

6. To add material to the model, right-click on the Materials node in the Model Builder and select Open Material Browser. In the Material Browser window, select Liquid and Gases>Liquids>Water and add it to the model by right-clicking on Water and selecting Add Material to Model.

Now we set up the boundary conditions and pipe shape and dimensions.

7. Click on the Pipe Properties 1 node in the Model Builder, located under Water Hammer (*whtd*). In the corresponding window, locate Pipe Shape section and select **Round** from the list. Enter **2*R** for Inner diameter. In the Pipe Model section, select User defined for Young's modulus and enter **E**. Similarly, enter **w** for Wall thickness. In the Flow Resistance section, select User defined for Friction model. Leave Darcy friction factor as zero. We accept this since the pipe's wall friction has minimal effects on the water hammer phenomenon, at least for most practical cases and during early cycles of oscillations. See Figure 4.127.

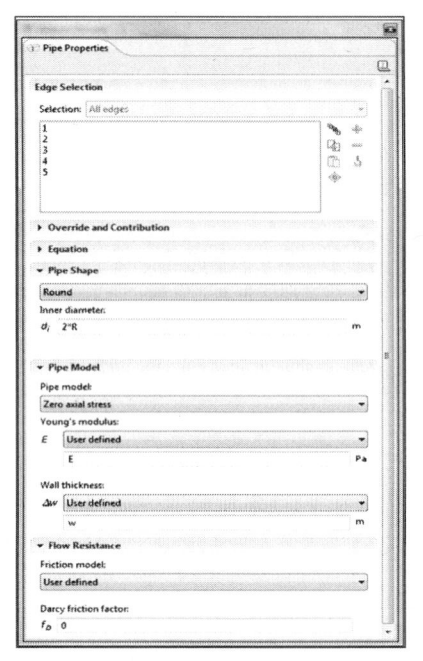

FIGURE 4.127 Pipe properties and shape.

8. Right-click on the Water Hammer (*whtd*) node in the Model Builder window and select Pressure from the list. In the Pressure window, add point 1 to the Selection list by clicking on the point at the start of the pipe network and then right-click. Enter **p0** for Pressure. Right-click on Water Hammer (*whtd*) node in the Model Builder window and select Local Friction Loss. In the corresponding window, add the points 3 and 4 representing the pipe elbow bends. Enter 0.9 for Loss coefficient (K_f for 90° bend). In the same window, expand the Constraint Settings section and check the box for *Use weak constraints*. This will help convergence of the solution for friction locations, like the bends in this example. (If Constraint Settings is not listed, click on the Show icon located in the toolbar of the Model Builder window and select Advanced Physics Options.)

9. Click on the Initial Values 1 node in the Model Builder window and in the corresponding window, under Initial Values section, enter **87958 Pa** for Pressure and **3.85 m/s** for Tangential velocity. The numerical values can be calculated for the pipe flow with set pressure **p0** at the entrance (point A) and velocity **u0** at the exit (point F). Alternatively,

we could add a pipe flow to this model to calculate the pressure and velocity.

10. To build the mesh, click on the Mesh 1 node in the Model Builder window. In the Mesh window, select User-controlled mesh from the list under Sequence type. Click on the Size node under Mesh 1 and in the corresponding window select Custom, then expand the Element Size Parameters section to enter **L/N** for Maximum element size and **1 [mm]** for Minimum element size. Right-click on the Mesh 1 node and select More Operations>Edge. From the Graphics window, add pipe line sections (1, 2, 3, 4, 5) to the Selection list. Click the Build All icon to build the mesh. Refer to Figure 4.128 for mesh parameters.

11. To set up the study, click on the Step 1: Time Dependent node in the Model Builder window. In the corresponding window, enter **range(0,1e-3,0.25)** for the Times. This will set a range starting from 0 seconds up to 0.25 seconds, with a time step of 0.001 seconds for capturing and storing the solutions into the solution data base. The maximum run time can easily be estimated by multiples of the time required for the pressure wave to travel once over the length of the pipe line (about 25/1200 sec.).

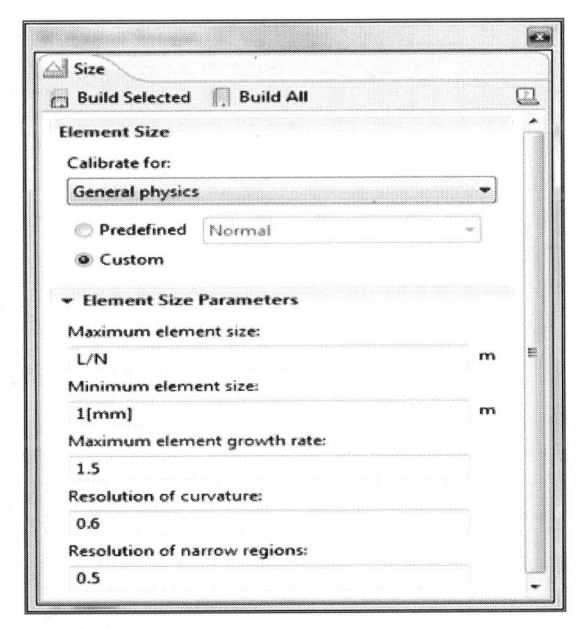

FIGURE 4.128 Custom mesh size and parameters.

12. For the solver time step, we use **dt**. Right-click on the Study 1 node in the Model Builder and select Show Default Solver from the list. A new Solver 1 node will appear in the model tree. Expand the Solver 1 node and click on Time-Dependent Solver 1. In the corresponding window, locate the Time Stepping section and check Maximum step to enter **dt**. For initial step, enter **0.5*dt**. See Figure 4.129.

13. Right-click on the Study 1 node and select Compute ≡. Wait for computations to finish.

14. For visualization of results, we use line graphs. Right-click on the Results node and select 1D Plot Group. Rename the newly generated line group to Pressure Line Plot. Right-click on the Pressure Line Plot node and select Line Graph. In the corresponding window, choose from the list option for Time selection, then click to highlight 0.244 from the values listed under Times. Expand the y-Axis Data section

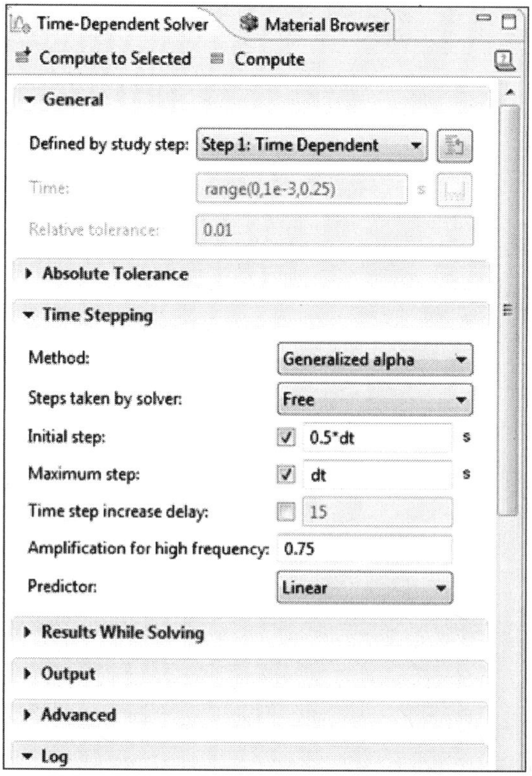

FIGURE 4.129 Transient solver parameters.

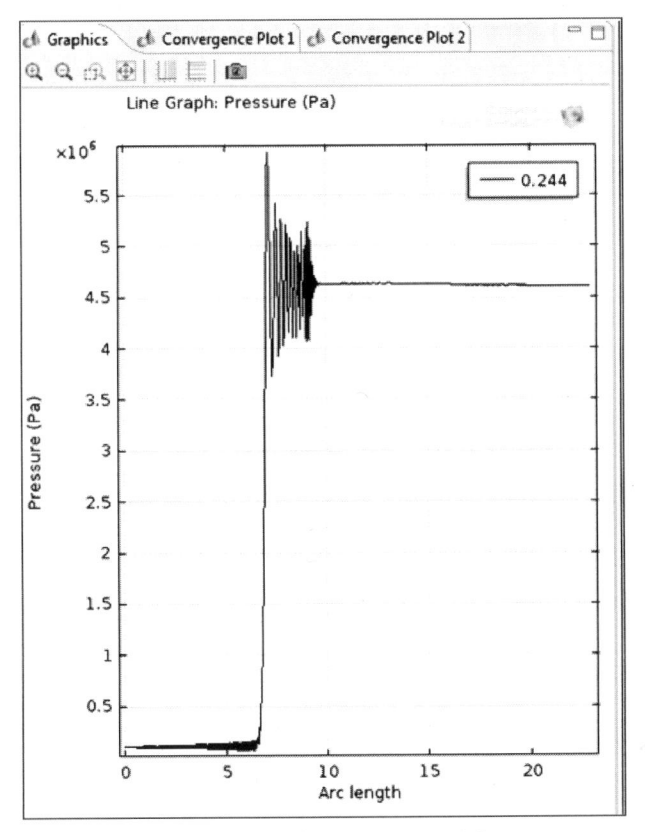

FIGURE 4.130 Pressure pulse for water hammer phenomenon at t = 0.244 s.

and enter **p** for Expression. Click Plot. The results for the water wave pressure at time 0.244 seconds appears in the Graphics window. See Figure 4.130.

15. Another useful graph is to draw the pressures at the gauges (measurement points 2 and 5). Right-click on the Results node and select 1D Plot Group. Rename the newly generated line group to Pressure point Plot. Right-click on the Pressure point Plot node and select Point Graph. In the corresponding window, select and add points 2 and 5 to the Selection list. In the y-Axis Data, type in **p.** Click Plot. The results appear in the Graphics window. Pressure variations for points 2 and 5 versus time clearly show the water wave moving back and forth through the pipe and measured at these points. Results are shown in Figure 4.131.

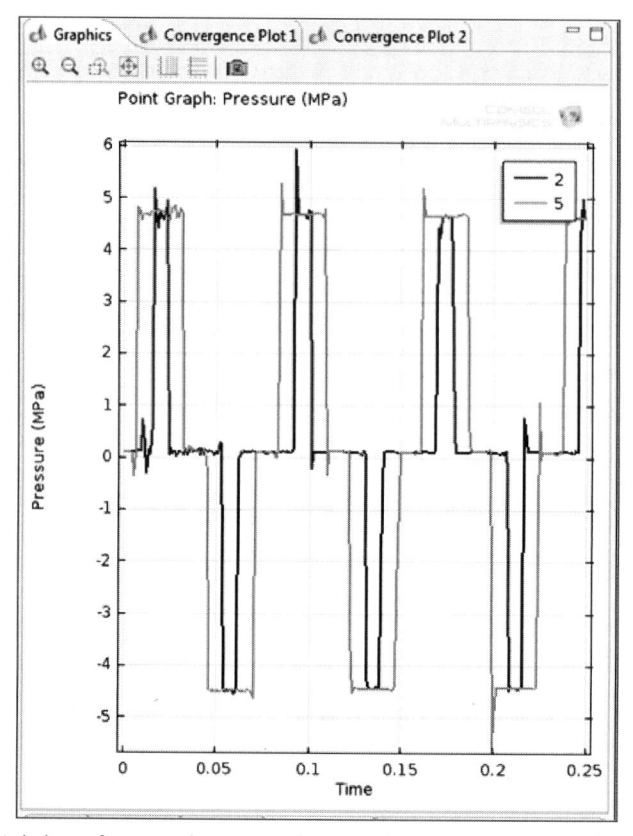

FIGURE 4.131 Variations of pressure due to water hammer phenomenon at gauge points 2 and 5.

Example 4.13: Static Fluid Mixer Model

In this example, we use flow analysis tools (Computational Fluid Dynamics, or CFD) in COMSOL 4.3 for analyzing the flow in a static mixer. The mixer is a 2D box mixer with dimensions 60 cm by 40 cm and has two inlets and one outlet. Three internal plates separate the internal space of the mixer to guide the flow for mixing. We build a model for this mixer using COMSOL.

Solution

1. Launch COMSOL, and open a new file. Save it as Example 4.13.
2. Select 2D from the Model Wizard window and click ⇨ Next. In the Add Physics window, expand the list for Fluid Flow and select Single-Phase Flow>Laminar Flow (*spf*). Click on the Laminar Flow node and

add it to the Selected physics list by clicking on the ✚ (or right-click on it and click Add selected). Then click on ➡ to go to the next window: Selected Study Type. In this window, click on the Stationary node and click on the flag icon 🏁 to Finish.

3. To build the geometry of the mixer:

   3.1 Click on the Geometry 1 node in the Model Builder window (if not already highlighted). In the Geometry window under Units, change the Length unit to **cm** from the list (open the list to see unit options). Draw a box by right-clicking on the Geometry 1 node and select Rectangle. The Rectangle window will open. In this window, enter 40 for Width and 60 for Height in the Size section, and 0 for x and 0 for y under Position. Make sure that Corner is selected for Base. Click on the Build Selected icon. The mixer box geometry will appear in the Graphics window, as shown in Figure 4.132.

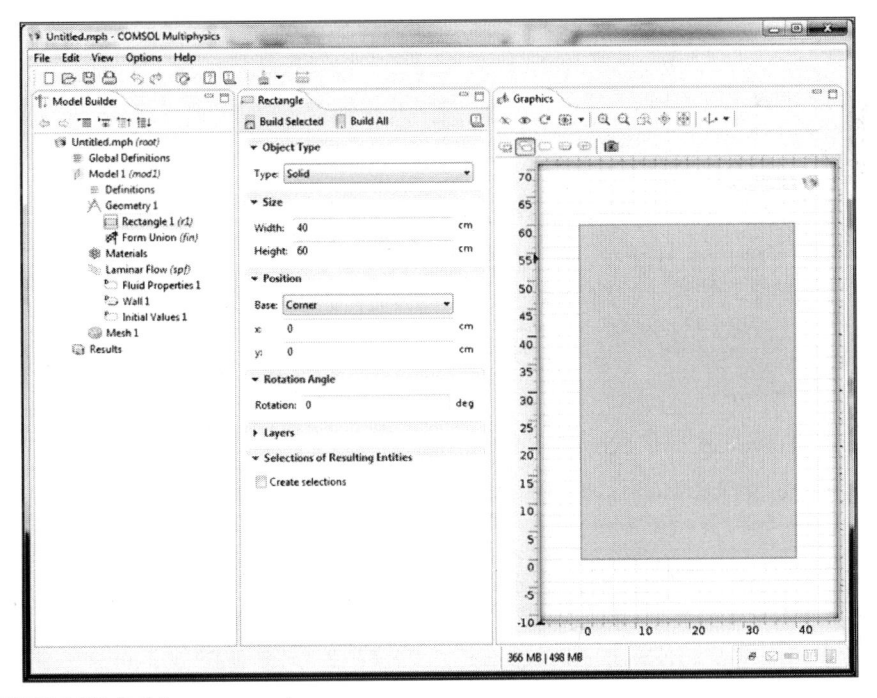

FIGURE 4.132 Building a rectangular geometry.

3.2 Similarly, draw three more boxes for inlets and out manifolds. Right-click on the Geometry 1 node in the Model Builder window and select Rectangle. In the Rectangle window, enter 6 for Width and 4 for Height in the Size section, and –6 for x and 52 for y under Position. Make sure that Corner is selected for Base. Click on the Build Selected icon . Again, right-click on the Geometry 1 node in the Model Builder window and select Rectangle. In the Rectangle window, enter 6 for Width and 4 for Height in the Size section, enter –6 for x and 4 for y under Position, and make sure that Corner is selected for Base. Click on . Finally, right-click on the Geometry 1 node in the Model Builder window and select Rectangle. In the Rectangle window, enter 12 for Width and 6 for Height in the Size section, enter 40 for x and 26 for y under Position, and select Corner for Base. Click on . Results are shown in Figure 4.133.

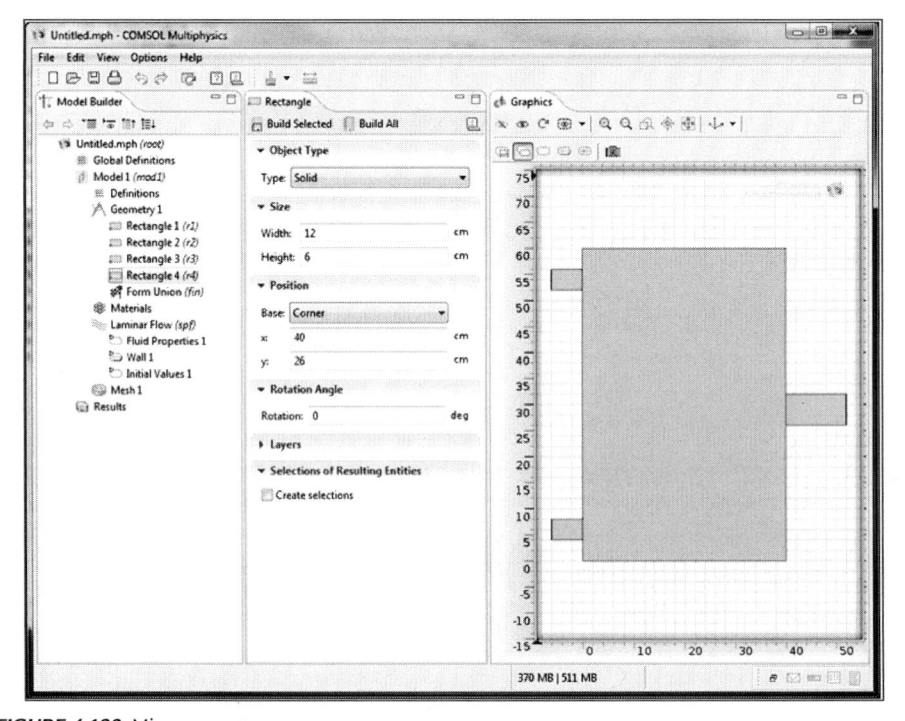

FIGURE 4.133 Mixer geometry.

3.3 Next, draw the baffles. Click on the Geometry 1 node in the Model Builder. Click on the Draw Line icon in the toolbar and draw a vertical line inside the mixer box (click on a point inside the box, then move the mouse and click again; to release, right-click). In the Bezier Polygon window, click on Segment 1 (*linear*) located in the area under the Added segments. In the Control points section, enter the coordinates of the line points as (12, 60) and (12, 40) and then click the Build Selected icon. The line will be located in the right position in the Graphics window. Similarly, draw two more lines with coordinates as {(12, 0), (12, 20)} and {(26, 50), (26, 10)}. The final Geometry should be as shown in Figure 4.134.

4. To assign the fluid properties, right-click on the Materials node in the Model Builder and select Open Material Browser. In the Material Browser window, expand the Liquid and Gases>Liquids node, right-click on Engine Oil, and select Add Material to Model. The Material window will open and four domains (mixer, two inlets, and one outlet) will be listed in the Selection area. Note that the Dynamic viscosity and density of the fluid is checked with ✓. In the Value column under

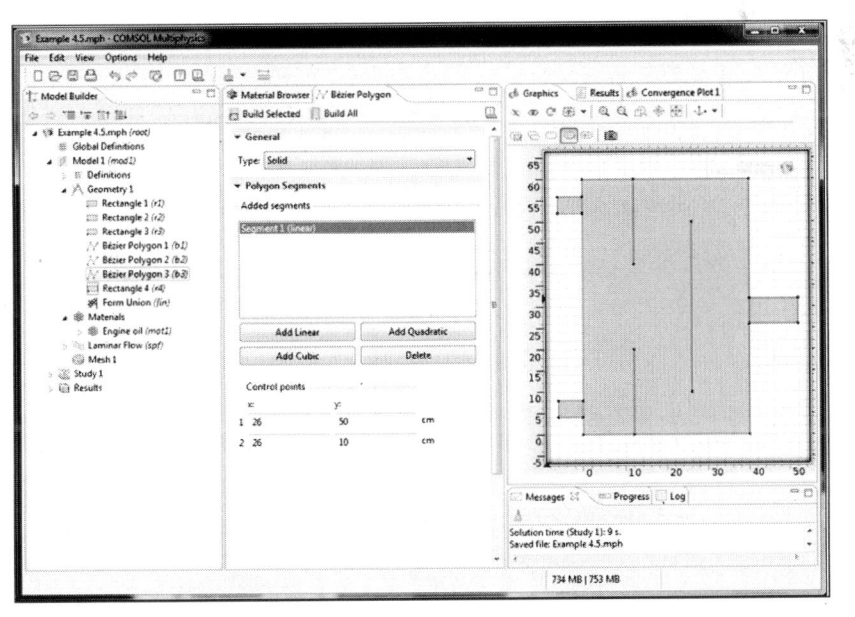

FIGURE 4.134 Mixer geometry with internal baffles.

Material Contents section, these properties are given as a function of T, temperature. Therefore, we should define the value of temperature for which we would like to calculate these properties. To do this, right-click on the Global Definitions node in the Model Builder window and select Parameters. The Parameters window will open. In this window, type **T** under Name and **40[degC]** under Expression. See Figure 4.135.

5. To assign the boundary conditions:

    5.1. For inlets, right-click on the Laminar Flow (*spf*) node in the Model Builder window and select Inlet. The Inlet window will open. In this window, add the entrance edges of the inlets to the Boundary Selection list by clicking on the corresponding edges of the mixer inlets in the Graphics window and click on the plus icon ✚. In the Velocity section, choose Normal inflow velocity and enter a value of 1 for U_0.

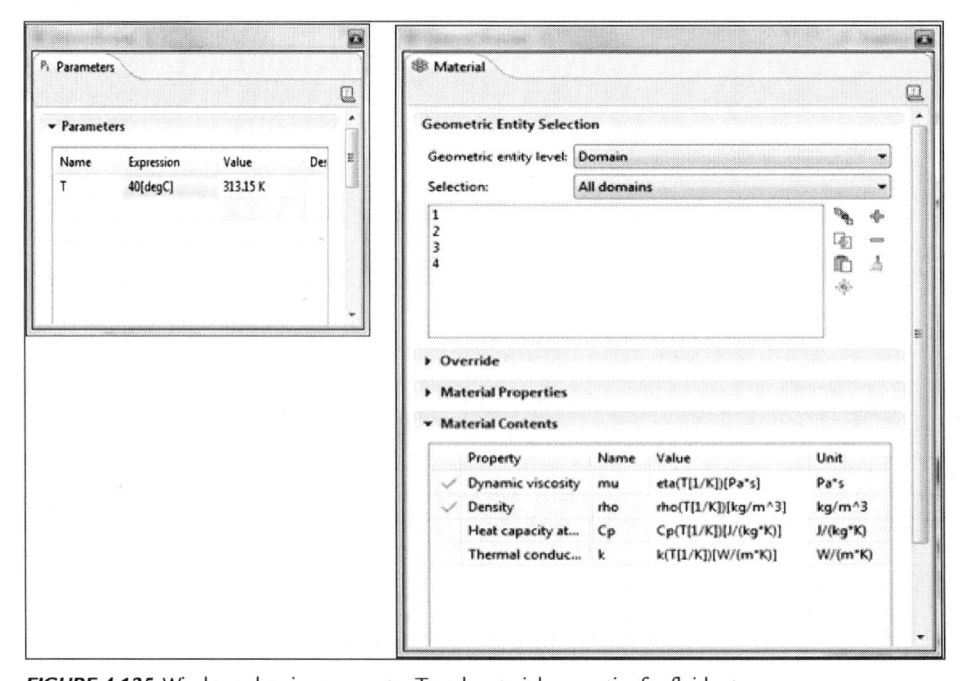

FIGURE 4.135 Windows showing parameter T and material properties for fluid setup.

5.2. Similarly, assign the boundary conditions for the outlet. Right-click on the Laminar Flow (*spf*) node in the Model Builder window and select Outlet. The Outlet window will open. In this window, add the edge of the outlet to the Boundary Selection list by clicking on the corresponding edge of the mixer outlet in the Graphics window and click on the plus icon ✚. In the Boundary Condition section, choose Pressure, no viscous stress option from the list (if not already selected by default) and enter 0 for value of p_0. Assuming zero pressure for the atmospheric pressure gives the pressure field inside the mixer as gauge pressure.

5.3. We should define the baffles as no-slip internal boundaries. Right-click on the Laminar Flow (*spf*) and select Interior Wall. The Interior Wall window will open. In this window, add the three lines/baffles by clicking on them in the Graphics window and click on the ✚. Make sure No slip option is selected in the Boundary condition section. See Figure 4.136.

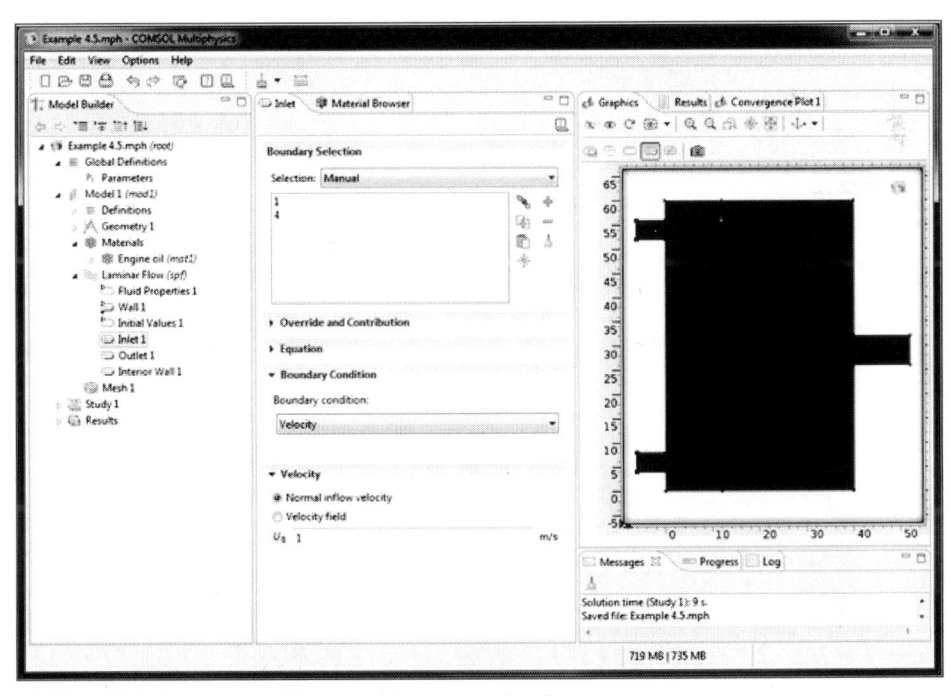

FIGURE 4.136 Boundary conditions setup for inlets and outlet.

6. For creating a mesh, click on the Mesh 1 node in the Model Builder, and in the Mesh window select Coarse for Element size under Mesh Settings. Make sure the default sequence type Physics-controlled mesh is selected. Click the Build All icon ▣ to create the mesh. The mesh will appear in the Graphics window. Note that the mesh is a hybrid one that is a combination of quadrilateral boundary layer and triangular elements, as shown in Fig 4.137.

7. To run the model, right-click on the Study 1 node and select Compute ≡ . The default velocity field will appear in the Graphics window.

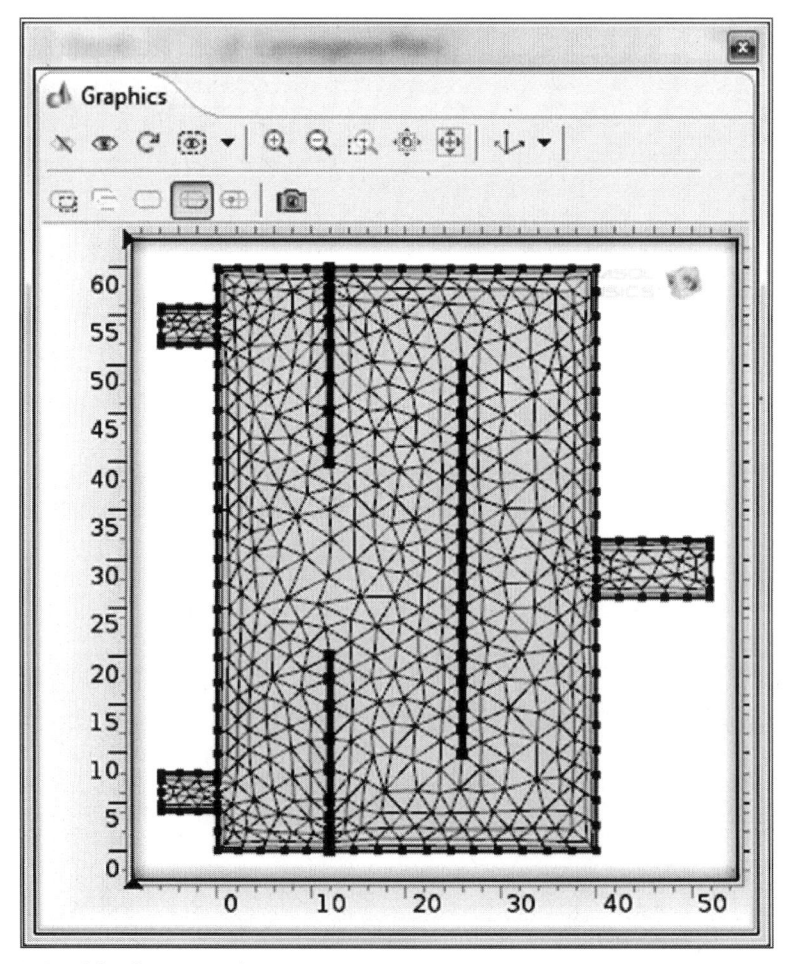

FIGURE 4.137 Finite elements mesh.

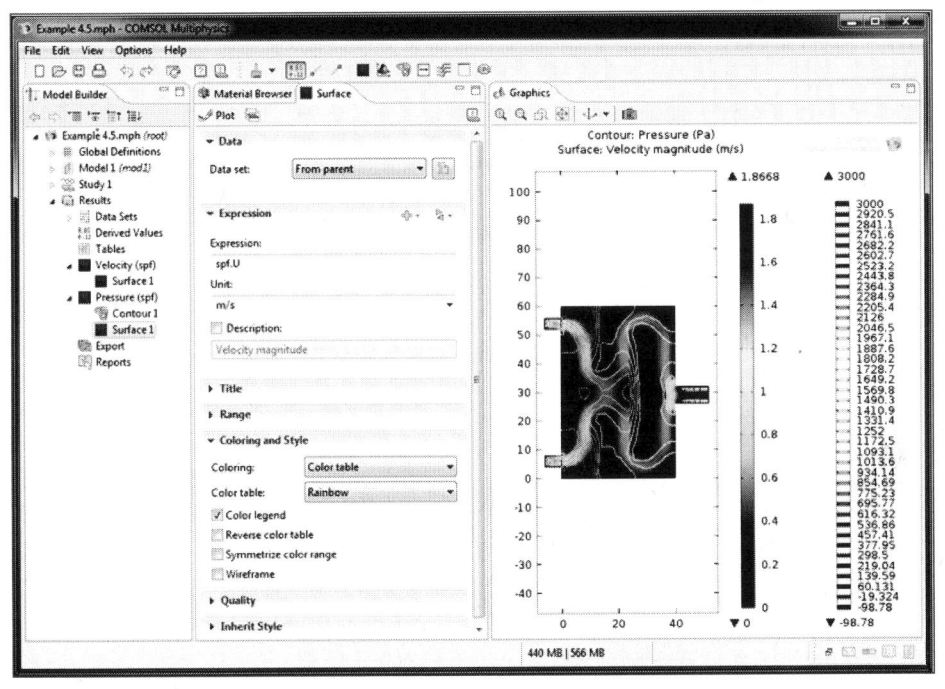

FIGURE 4.138 Results for flow velocity and pressure contours.

8. To manipulate the results, click on the Contour 1 under Pressure (*spf*) to show the pressure contours in the Graphics window. Then right-click on the Pressure (*spf*) and select Surface. A Surface 1 node will appear under Contour 1. In the Graphics window, the velocity field will appear mapped on the pressure contour, as shown in Figure 4.138.

9. It would be useful to calculate the Reynolds number for fluid flow, which helps to confirm the assumption of laminar flow made for this analysis. In COMSOL, the cell Reynolds number is defined as $Rec = \rho|u|h/(2\mu)$, where $|u|$ is the velocity magnitude, h is the finite element length, ρ fluid density, and μ fluid dynamic viscosity. Click on the Surface 1 under Velocity (*spf*) node, in the Surface window click Replace Expression, and click on Laminar Flow>Cell Reynolds number (*spf.cellRe*). The values of the Reynolds number will appear in the Graphics window. The maximum value is 12.996. Note that this is a cell Reynolds number and its value depends on the resolution of the mesh used.

Therefore, if we build a mesh with smaller element size the Reynolds number will be smaller, as well. This is typical for Reynolds number calculation and for any specific problem we need to identify the characteristic length scale based on which the Reynolds number is defined. For this example, a coarse mesh has a maximum element size of about 5 cm, which is close to the outlet width (about 6 cm) where the maximum velocity is. Therefore, we can accept a Reynolds number of about 13. See Figure 4.139.

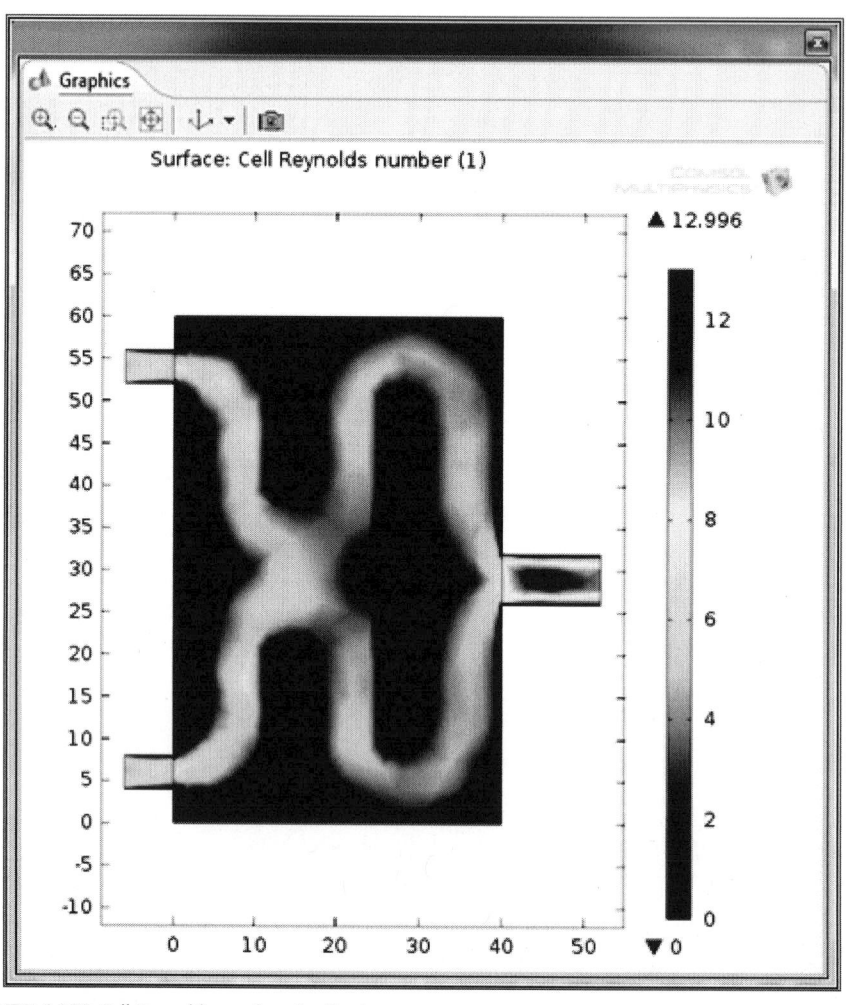

FIGURE 4.139 Cell Reynolds number distribution.

SECTION 4.3: MODELING OF STEADY AND UNSTEADY HEAT TRANSFER IN MEDIA

In this section, we use COMSOL modules to model some examples in heat transfer. Models include steady, transient conduction, and convection in two- and three-dimensional media. The main objective is to provide, for users and readers, some solved examples that can be used directly or lead to further solutions of similar or more complex structures using COMSOL. It is assumed that readers are familiar with relevant engineering principles and governing equations related to heat transfer (see Reference 4.8). In each example, we provide brief explanations of physics involved along with governing equations and phenomena, as applicable. It is recommended that readers cover previous Chapters 1 through 3 before attempting examples in this section.

Example 4.14: Heat Transfer in a Multilayer Sphere

In this example, we model a steady state heat transfer in a hollow multilayer sphere. The governing equation for heat transfer in solids is:

$$\nabla \cdot (k \nabla T) + Q = 0$$

where k is material thermal conductivity, T temperature, and Q heat source/sink per unit volume. The quantity $|k \nabla T|$ is the heat flux or thermal energy per unit area per unit time, according to Fourier's law. For the domain geometry, we assume a composite multilayer hollow sphere. We use the axisymmetric feature of the sphere for creating the geometry of the model and therefore solve a 3D problem as a 2D axisymmetric one. This approach saves on computation time and computer memory and resources.

Solution

1. Launch COMSOL and open a new file. Save the file as Example 4.14.

2. In the Model Wizard window, select 2D axisymmetric and click Next.

3. From the Add Physics list, expand Heat Transfer and select Heat Transfer in Solids (ht) and add it to the Selected physics list by right-clicking on it and choosing Add Selected. Then click Next.

4. From the Select Study Type, click on Stationary and then click on Finish.

To build the geometry, we create several half-circles using the parametric curves tool available in COMSOL.

5. Click on the Geometry node in the Model Builder window and select **cm** from the list for Length unit for Units. Then right-click on the Geometry node again and select Parametric Curve. In the corresponding window, enter the following data and click Build All to create a half-circle of radius 2 cm. See Figure 4.140.

6. Right-click on Parametric Curve 1 (*pc1*) and select Duplicate. A new node will appear. In the corresponding window, change the data under Expressions for r: to **3*cos(s)** and z: to **3*sin(s)**, only. Periodically click Zoom Extents in the Graphics window to see the entire geometry. Click Build All.

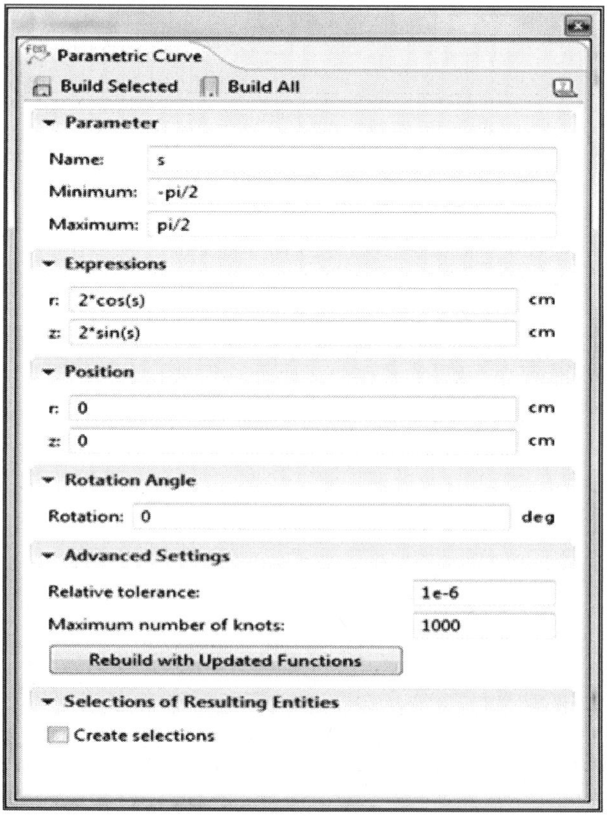

FIGURE 4.140 Parametric curve data for half-circle.

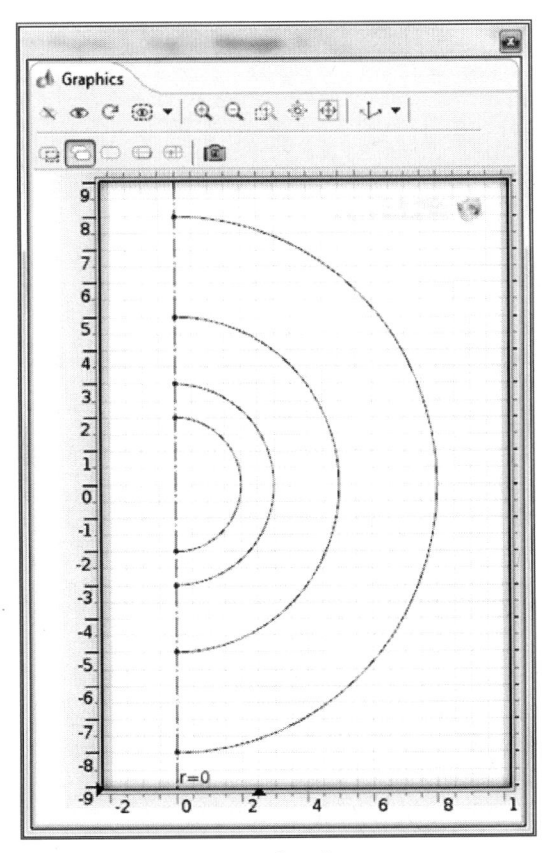

FIGURE 4.141 Resulting 2D axisymmetric geometry for spheres.

7. Repeat actions descibed in Step 6 to create two more half circles with radii **5 cm** and **8 cm.** See Figure 4.141.

8. To finish building the model geometry, we should create solids or domains by drawing several lines that connect the end points of the circles located on the **r = 0** axis. Click on the Geometry node and select Draw Line from the list in the main toolbar. Click on an end point and then the next node to draw a line; to release, right-click. Make sure that the two ends of the central half-circle are not connected, since we would like to have a hollow space at the center of the sphere. Finally, select all entities in the Graphics window (click at any point in the Graphics window and then press CTRL+A). While all entities are highlighted, click on the Convert to Solid icon located in the main toolbar, as shown in Figure 4.142.

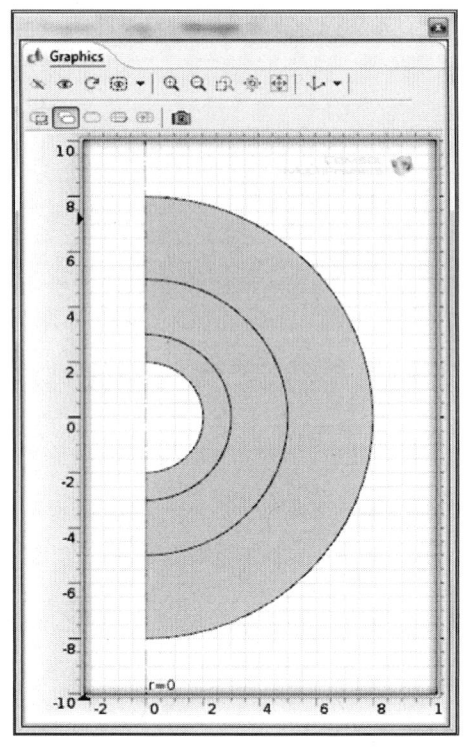

FIGURE 4.142 Resulting 2D axisymmetric domains.

9. To add materials to the model, right-click on the Materials node and select Open Material Browser. In the corresponding window, expand Built-in, select Cast iron from the list by right-clicking on it, and choose Add Material to Model. In the corresponding window, add the outer layer (layer 1, only) to the Selection list by clicking on this layer in the Graphics window and then right-click. Enter 23 for k and 506 for Cp in the Material Contents section (you may need to expand this section to see the list).

10. Repeat actions described for Step 9 to add material Aluminum 6063-T83 to the middle layer (layer 2) and silicon to the inner layer (layer 3).

11. To define the boundary conditions, we set a temperature at the inner surface and a convective cooling one at the outer surface. Right-click on the Heat Transfer in Solids (ht) node in the Model Builder window and select Temperature from the list. In the corresponding window,

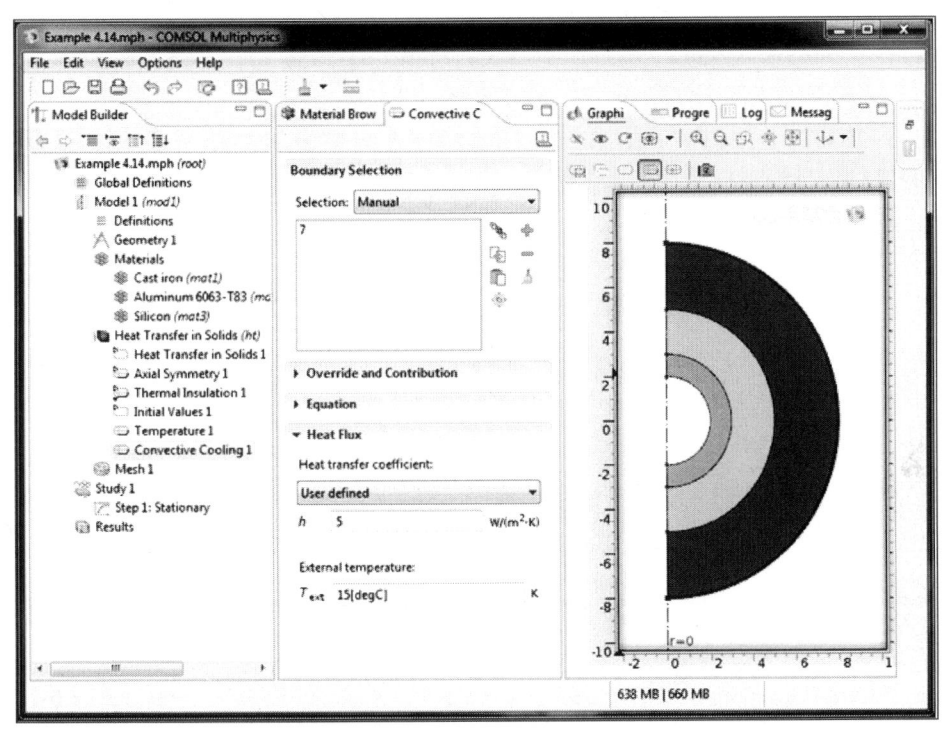

FIGURE 4.143 Materials and properties assigned to spherical layers.

add the inner boundary (number 10) to the Selection list in the Boundary Selection section. Enter 450[degC] for Temperature. Similarly, select Convective Cooling 1 and assign it to the outer boundary (number 7) in the Convective Cooling window. In this window in the Heat Flux section, enter 5 for heat transfer coefficient and 15[degC] for External temperature. See Figure 4.143.

12. To build a mesh, click on the Mesh 1 node. In the Mesh window, select Fine from the list for Element size and click Build All. A total of 464 finite elements will be created, as shown in Figure 4.144.

13. To run the model, right-click on the Study 1 node and select Compute. Wait for the computations to finish. The default results will appear in the Graphics window for the temperature values. Expand the Temperature, 3D (*ht*) node and click on Surface 1. In the corresponding window, change Unit to degC in the Expression section and Color table to Rainbow in the Coloring and Style section. See Figure 4.145.

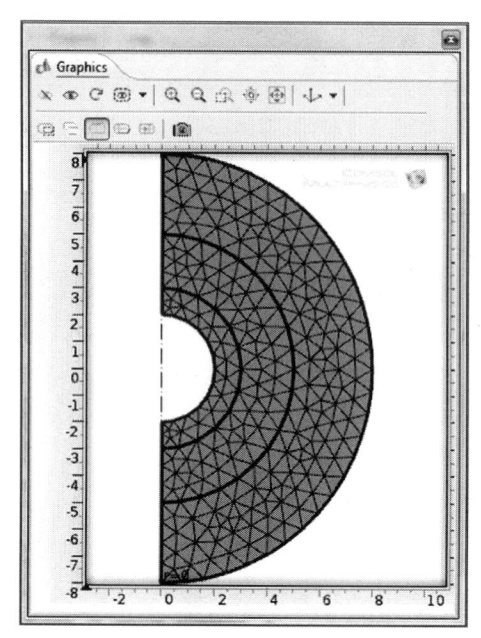

FIGURE 4.144 Finite elements mesh for spherical layers.

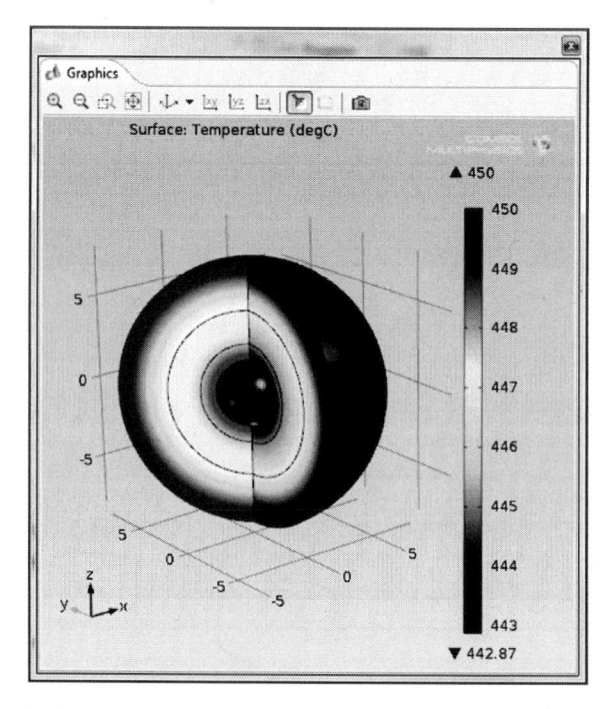

FIGURE 4.145 Results for temperature distribution.

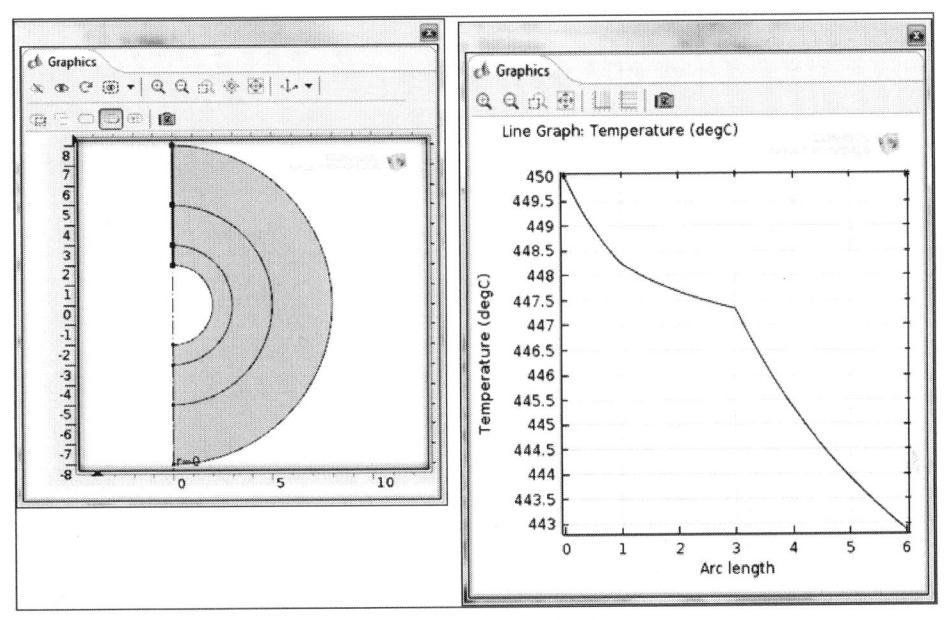

FIGURE 4.146 Windows showing results for temperature distribution across layers.

14. To draw the temperature on a graph, right-click on the Results node in the Model Builder and select 1D Plot Group. A new 1D plot will appear; right-click on it and select Line Graph. In the corresponding window for Line Graph, we can draw a line to plot the value of a quantity from the model. In this example, any radial line can be selected for this purpose. For example, select all boundaries along the upper part of the axis of symmetry (boundaries 4, 5, 6) from the Graphics window and add them to the Selection list (click on each boundary and then right-click). In the y-axis, select **degC** from the list for Unit. For the variable, select temperature T. Click the Plot icon. See Figure 4.146.

Example 4.15: Heat Transfer in a Hexagonal Fin

Temperature variation and heat transfer calculations in fins are needed for design of many heat transfer systems, such as heat exchangers. In this example, steady state heat transfer equations are solved for a hexagonal shape fin, as shown in Figure 4.147, including convective cooling boundary conditions.

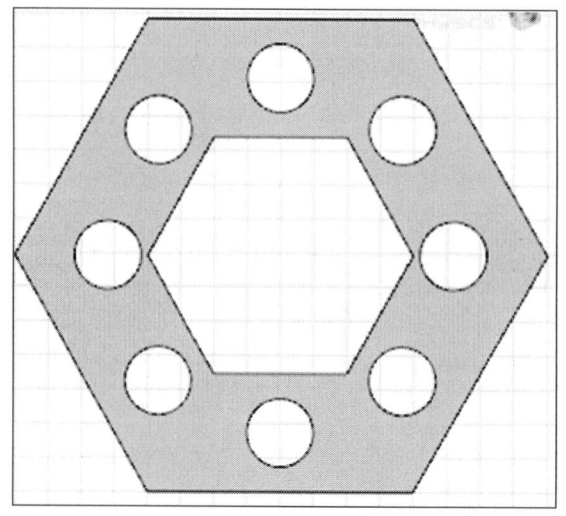

FIGURE 4.147 Geometry of the hexagonal fin.

Solution

1. Launch COMSOL and open a new file. Save the file as Example 4.15.
2. In the Model Wizard window, select 2D and click Next ⇨.
3. From the Add Physics list, expand Heat Transfer, select Heat Transfer in Solids (*ht*), and add it to the Selected physics list by right-clicking on it and choosing Add Selected. Then click Next ⇨.
4. From the Select Study Type, click on Stationary and then click on Finish.
5. In the Geometry window, select **cm** from the list for Length unit.
6. To draw a hexagonal, right-click on Geometry 1 in the Model Builder window and select Polygon from the list. In the Polygon window, enter the data as shown in Figure 4.148 for Coordinates of polygon vertices.
7. Similarly, create another polygon and enter the data as shown in Figure 4.149 for its Coordinates.
8. To subtract the second polygon from the first one, right-click on the Geometry 1 node and select Boolean Operations>Difference. In the Difference window, specify the first polygon as the Object to add, and the second polygon as the Object to subtract. Then click Build All.

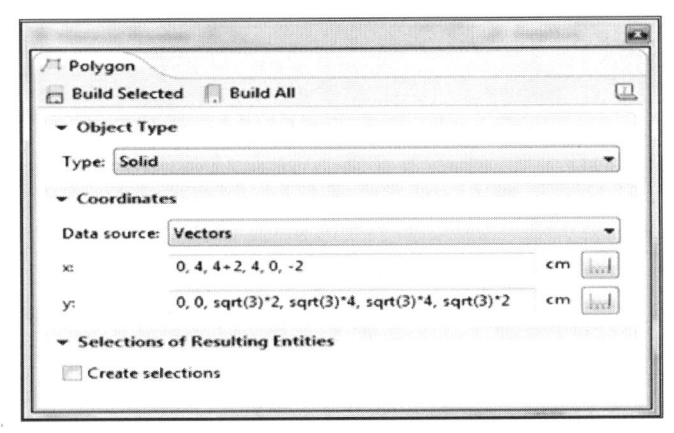

FIGURE 4.148 Polygon parameters for building outer hexagonal fin geometry.

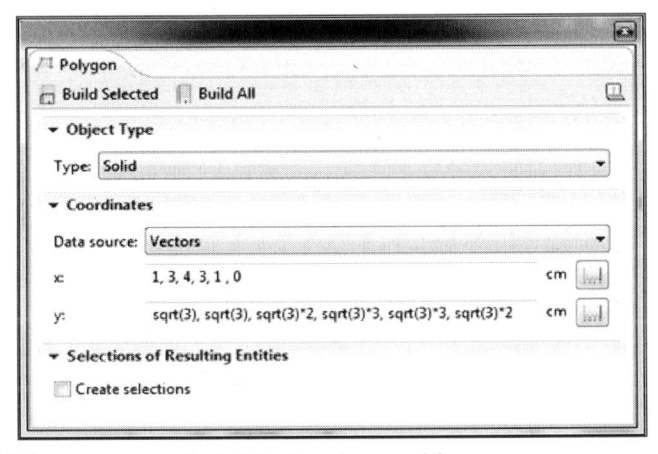

FIGURE 4.149 Polygon parameters for building inner hexagonal fin geometry.

9. To draw a circle, right-click on the Geometry 1 window and select Circle. In the corresponding window, enter the data as shown in Figure 4.150 for Radius and Center coordinates of the circle. Click Build All.

10. To create the other seven circles, right-click on the Geometry 1 node and select Transforms>Rotate. In the Rotate window, enter **range(0,45,315)** for Rotation Angle and enter **x:2** and **y:2*sqrt(3)** for Center of Rotation coordinates. Click Build All. See Figure 4.151.

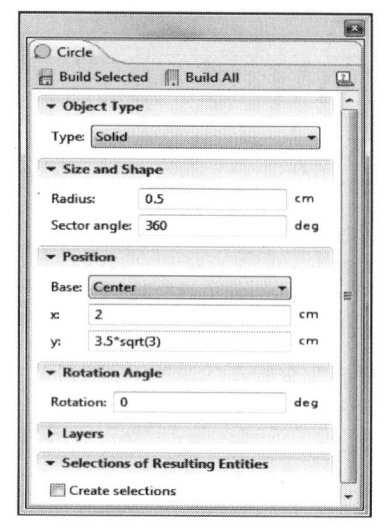

FIGURE 4.150 Parameters for building circular hole geometry.

FIGURE 4.151 Parameters for building Rotate circular holes geometry.

  11. Similar to Step 8 above, subtract the circles from the fin geometry. This time we use a different method. Click anywhere in the Graphics window and then press CTRL+A. All the geometry entities will change color to indicate they are selected. Then click the Difference icon located in the main toolbar. The final fin shape will appear in the Graphics window, as shown in Figure 4.152.

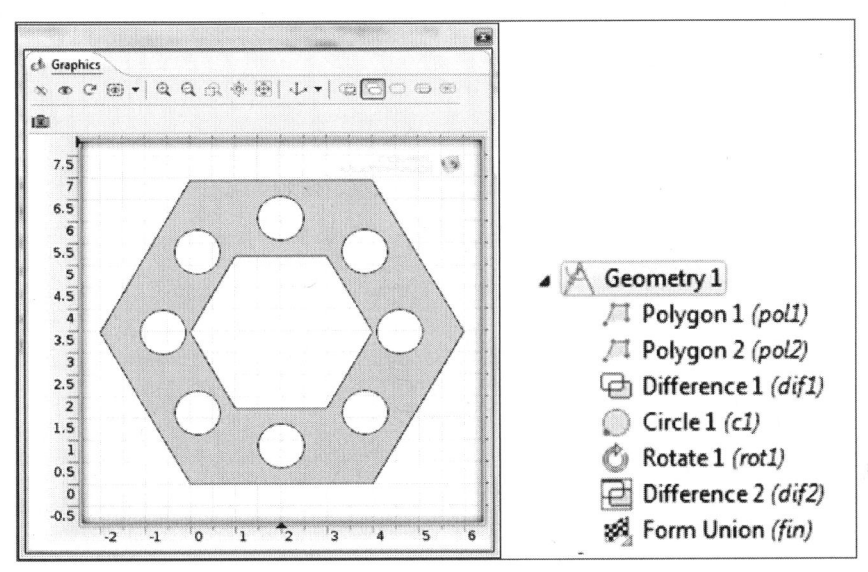

FIGURE 4.152 Resulting fin geometry and tree.

12. To add materials to the model, right-click on the Materials node in the Model Builder and select Open Material Browser. In the corresponding window, expand Built-In and select Aluminum. Add it to the model by clicking on the plus icon ✚ and selecting Add Material to Model.

13. Next, we add the boundary conditions to the model. Right-click on the Heat Transfer in Solids (ht) node in the model builder and select Convective Cooling. In the corresponding window, add all the outer edges (1, 2, 3, 6, 11, 12) of the fin to the Selection list. To do this, click on each edge in the Graphics window and then right-click. Expand the Heat Flux section and enter the Heat transfer coefficient **5** and External temperature **18[degC]**, as shown in Figure 4.153.

14. Similarly, create another Convective Cooling node and assign the inner edges of the fin to the Selection list. Enter the data as shown in Figure 4.154.

15. We assume four pipes passing through the fin circular holes with surface temperature of 250° and another four pipes with surface temperature of 80°. To assign these boundary conditions, right-click on the Heat Transfer in Solids (ht) node in the Model Builder and select Temperature. In the corresponding window, add the boundary

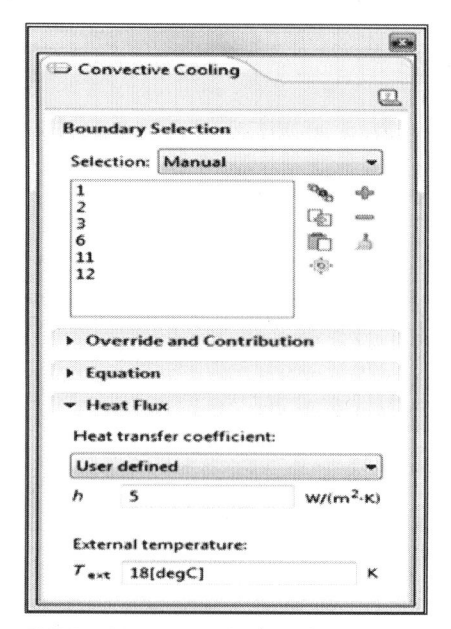

FIGURE 4.153 Convective boundary condition setup for outer edges.

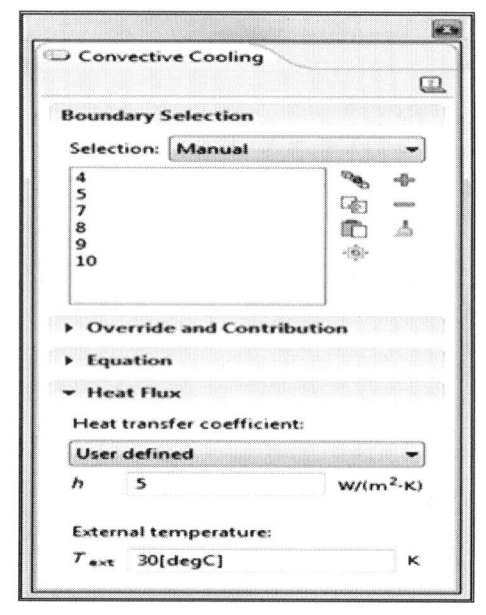

FIGURE 4.154 Convective boundary condition setup for inner edges.

of the circular holes to the Selection list, as shown in the left window in Figure 4.155. For Temperature, enter **250[degC]**. Similarly, create another Temperature node, assign the remaining circular holes boundaries to it, and enter **80[degC]** for Temperature, as shown in the right window in Figure 4.155

16. To build a mesh, click on the Mesh 1 node and select Fine for Element size in the Mesh window. Click Build All . Results are shown in Figure 4.156.

17. To run the model, right-click on Study 1 and select Compute . The default results will appear in the Graphics window after computations finish, as shown in Figure 4.157.

18. To show the direction of heat flux vector, right-click on the Isothermal Contours (ht) node and select Arrow Surface. In the corresponding window, replace x component data with **ht.tfluxx,** which represents the x-component of conductive heat flux, and replace y component: data with **ht.tfluxy.** Other parameters can be changed as desired. Click Plot . See Figure 4.158.

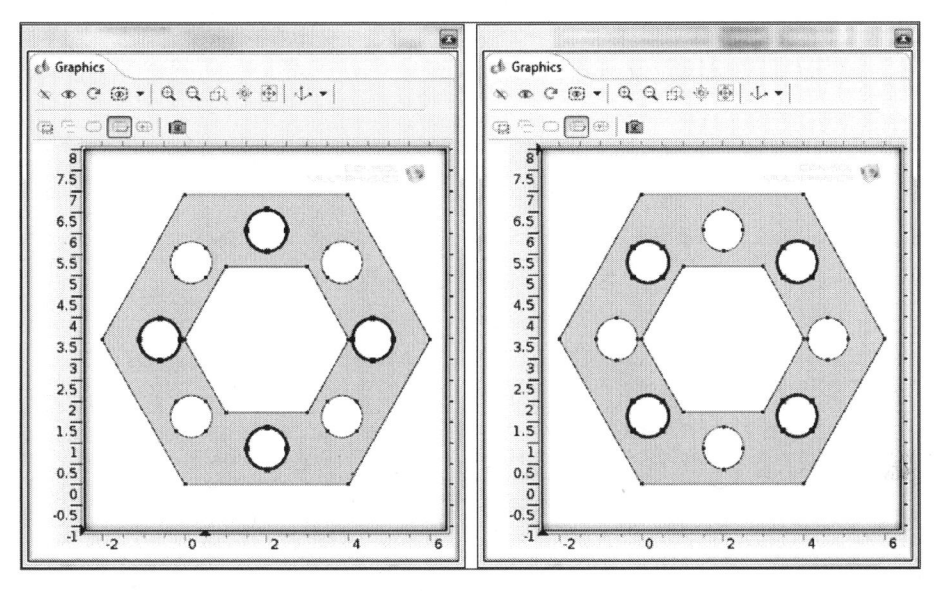

FIGURE 4.155 Windows showing temperature boundary condition setup for circular holes.

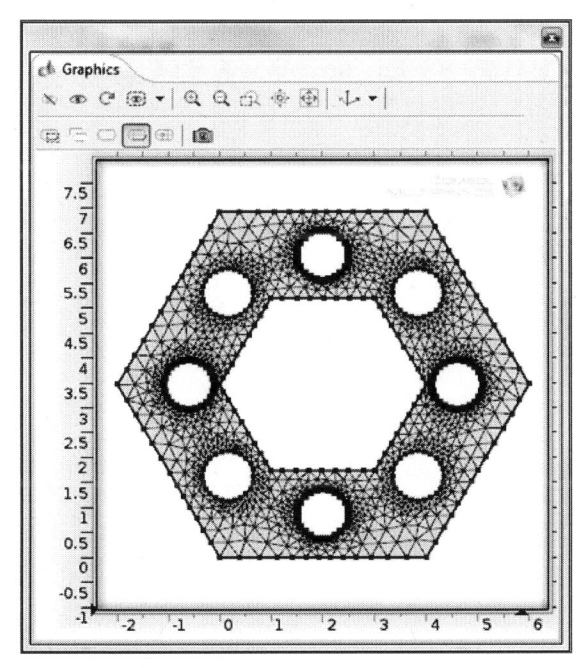

FIGURE 4.156 Finite elements mesh for the fin.

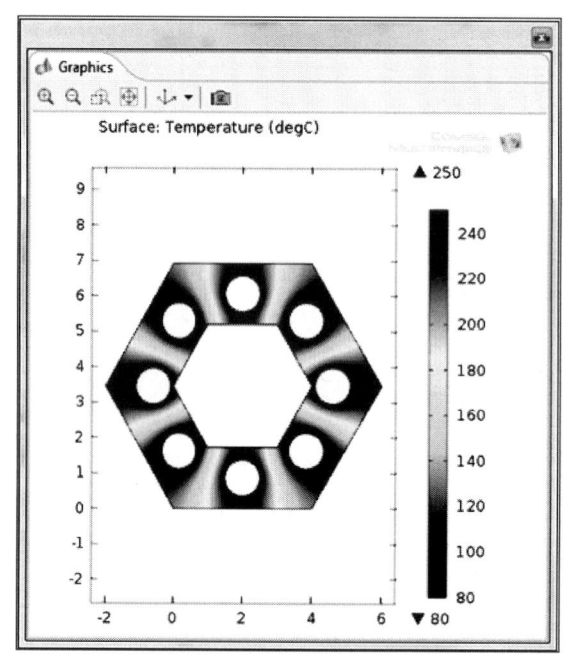

FIGURE 4.157 Results for temperature distribution for the fin.

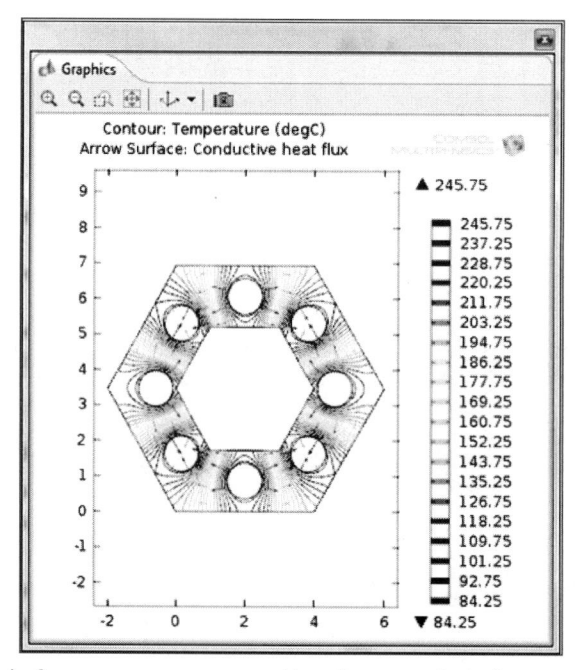

FIGURE 4.158 Results for temperature contours and heat flux vectors in the fin.

Example 4.16: Transient Heat Transfer Through a Nonprismatic Fin with Convective Cooling

In this example, we model transient heat transfer and variation of temperature through the body of a nonprismatic fin. For heat transfer modes, we consider heat conduction inside the fin material and heat convection from its surface to the ambient. The governing PDE for transient heat transfer is given as:

$$\rho C \frac{\partial T}{\partial t} = \nabla \cdot (k \nabla T) + Q$$

where ρ is material density, C heat capacity, k thermal conductivity, T temperature, and Q heat source/sink per unit volume. Two dimensionless numbers—Biot number ($Bi = hL/k$) and Fourier number ($F_o = \alpha t/L^2$)—are used for scaling the unsteady heat transfer and nondimensionalizing the governing equation. In these equations α is thermal diffusivity ($k/\rho C$, ratio of how much heat a material conducts versus how much it stores), h is heat transfer coefficient, and L is a length scale (usually, the ratio of volume of the heat transfer domain over the surface where heat convection occurs). For example, for a sphere L is equal to one-third of its radius.

The physical meanings of Bi and F_o are useful when dealing with modeling transient heat transfer problems.

$$Bi = \frac{\text{convection at the surface of the body}}{\text{conduction inside the body}}$$

$$= \frac{\text{conduction resistance inside the body}}{\text{convection resistance at the surface of body}}$$

Therefore, when $Bi < 1$, conductive heat transfer is faster than convection and the spatial variation of temperature inside a solid body can be neglected, as if whole body temperature changes only with respect to time. When $Bi > 1$, conduction resistance inside the body is larger than that of convection at its surface. For the latter case, spatial variation of temperature is considerable as compared to its temporal variation inside the body, and both should be included in the model. The Fourier number is basically a dimensionless time scale defined as:

$$Fo = \frac{\text{rate of conduction}}{\text{rate of heat storage}}$$

For modeling an unsteady heat transfer, it is crucial to have the right time step/scale for numerical computation. If time scale is too large (versus characteristic time scale of the problem), then transient variation cannot be modeled/captured. Whereas if time step is too small, then unrealistic temperature fluctuation may show up in modeling results. When Bi is small (usually, smaller than 0.1), then we can neglect the spatial variation of temperature inside the body. In other words, for small Bi the body temperature changes mainly with respect to time. The product $B_i \times F_o$ can be used to estimate characteristic time.

$$Bi \times Fo = \left(\frac{hL}{k}\right)\left(\frac{\alpha \Delta t}{L^2}\right) = \left(\frac{h}{\rho CL}\right)\Delta t = b \Delta t$$

The inverse of b (or $\rho CL/h$) is the characteristic time. In other words, a large value of b indicates that temperature reaches the ambient temperature in a short amount of time (see Reference 4.8).

For this example, we consider a fin attached to a section of a wall, as shown in Figure 4.159.

Solution

1. Launch COMSOL and open a new file. Save the file as Example 4.16.
2. In the Model Wizard window, select 3D and click Next ⇨.

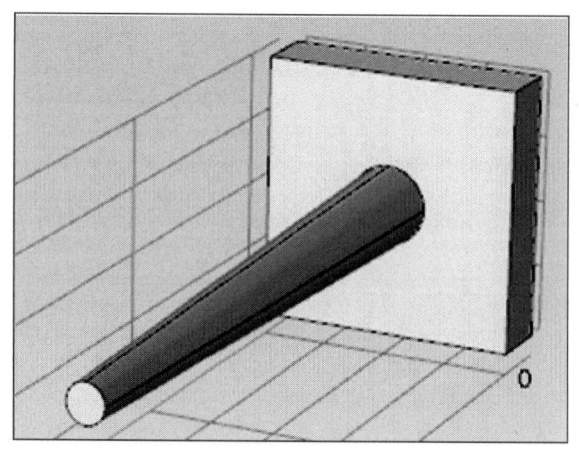

FIGURE 4.159 Fin geometry attached to a section of wall.

3. From the Add Physics list, expand Heat Transfer, select Heat Transfer in Solids (*ht*), and add it to the Selected physics list by right-clicking on it and choosing Add Selected. Then click Next ⇨.

4. Under Select Study Type, click on Time Dependent and then click on Finish. In the Geometry window, select **mm** from the Length unit list.

5. To build the geometry, right-click the Geometry 1 node and select Block from the list. Enter the data shown in Figure 4.160 for Width, Depth, and Height.

6. Again, right-click the Geometry 1 node and select Cone. Enter the data shown in Figure 4.161, and then click Build All.

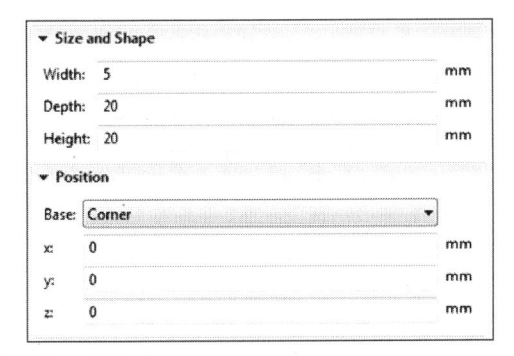

FIGURE 4.160 dimensions for wall block geometry.

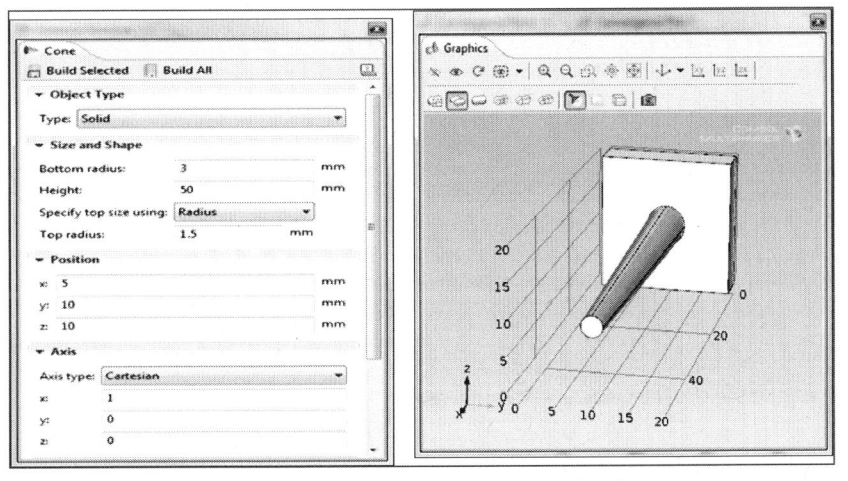

FIGURE 4.161 Fin cone parameters setup and final geometry of fin and wall.

7. Rename Block 1 to Wall and Cone 1 to Fin.

8. To add material to the model, right-click on Materials node in the Model Builder window and select Open Material Browser. In the Corresponding window, expand Built-In and select and add Aluminum to the model by clicking on ✚. Notice that Aluminum material is added to both the wall and fin (Domains 1 & 2). If needed, different materials can be assigned to each element. See Figure 4.162.

9. To define the boundary conditions, right-click on Heat Transfer in Solids (*ht*) and select Temperature from the list. In the Temperature window, select the face of the wall located at y-z plane (Boundary 1) from the Graphics window and add it to the Selection list. For Temperature, enter **450[degC]**.

10. Right-click on Heat Transfer in Solids (*ht*) and select Convective Cooling. In the corresponding window, select all surfaces of the fin and only the front face of the wall (Boundaries 6, 8, 9, 10, 11, 12). In the Heat Flux section, select User defined and enter 50 for Heat transfer coefficient and 30[degC] for External temperature. See Figure 4.163.

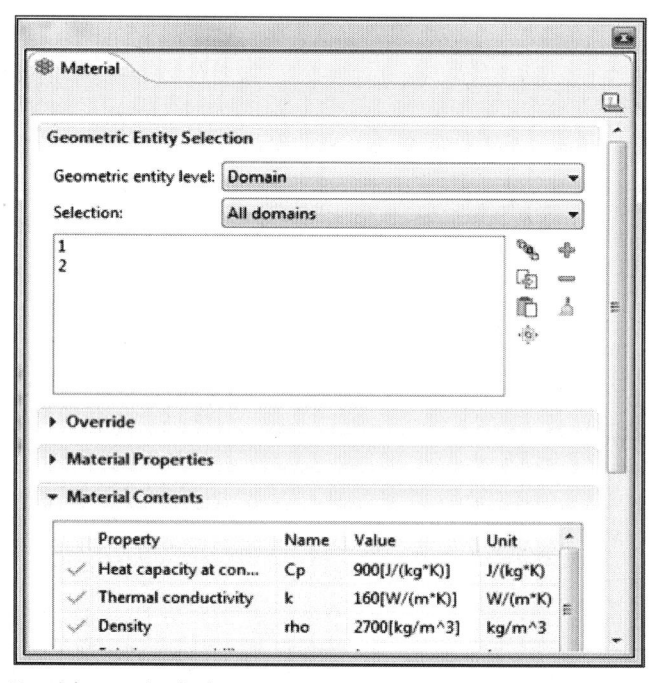

FIGURE 4.162 Material properties for fin.

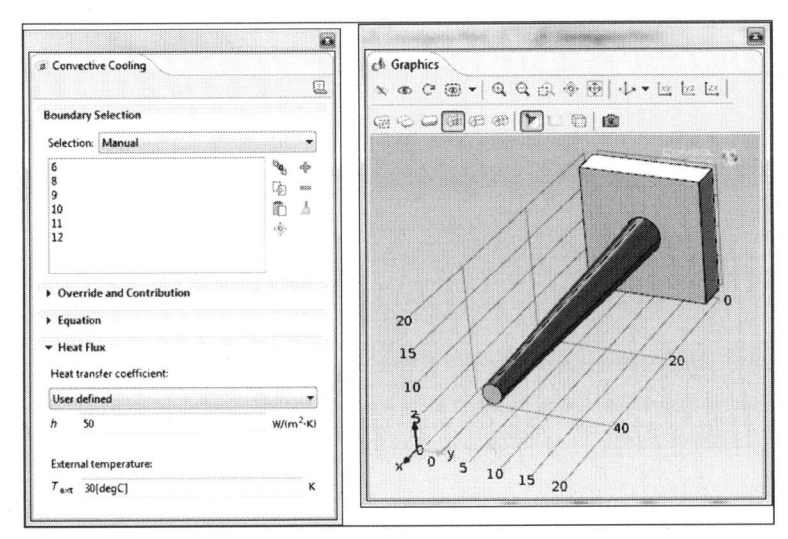

FIGURE 4.163 Convective boundary condition set up for fin.

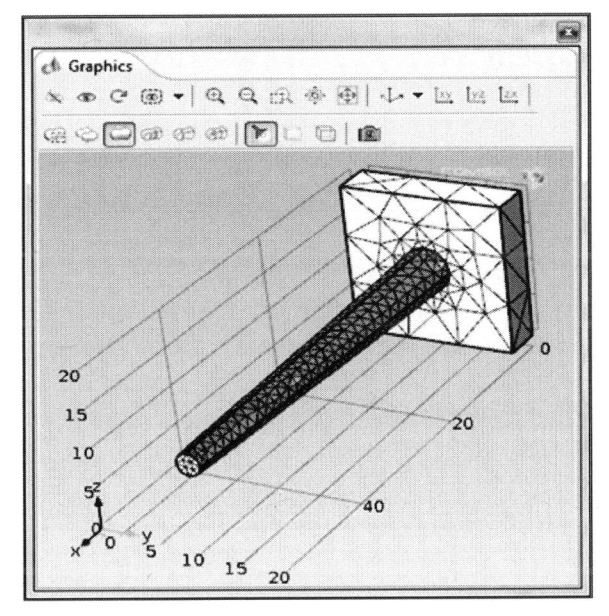

FIGURE 4.164 Finite elements mesh for fin and wall section.

11. To build a mesh, click on the Mesh 1 node. In the corresponding window, select Coarse for Element size. Click Build All . Mesh consists of total 1878 elements, as shown in Figure 4.164.

The value of b for this problem is 1.83E-2, which corresponds to a time constant 54.675 seconds. Therefore, it takes about 55 seconds for temperature to reach its equilibrium (i.e., temperature would not vary versus time at a given point). We use a small time step of 0.5 second for the first 5 seconds and a 2-second time step for the remaining times, up to 60 seconds.

12. To run the model, expand the Study 1 node and click on Step 1: Time Dependent node. In the corresponding window, enter **range(0, 0.5, 5), range(6, 2, 60)** for Times. Then right-click on Study 1 and select Compute. Default results for temperature will appear in the Graphics window, after calculations are complete. See Figure 4.165 for results at t = 30 sec., as an example.

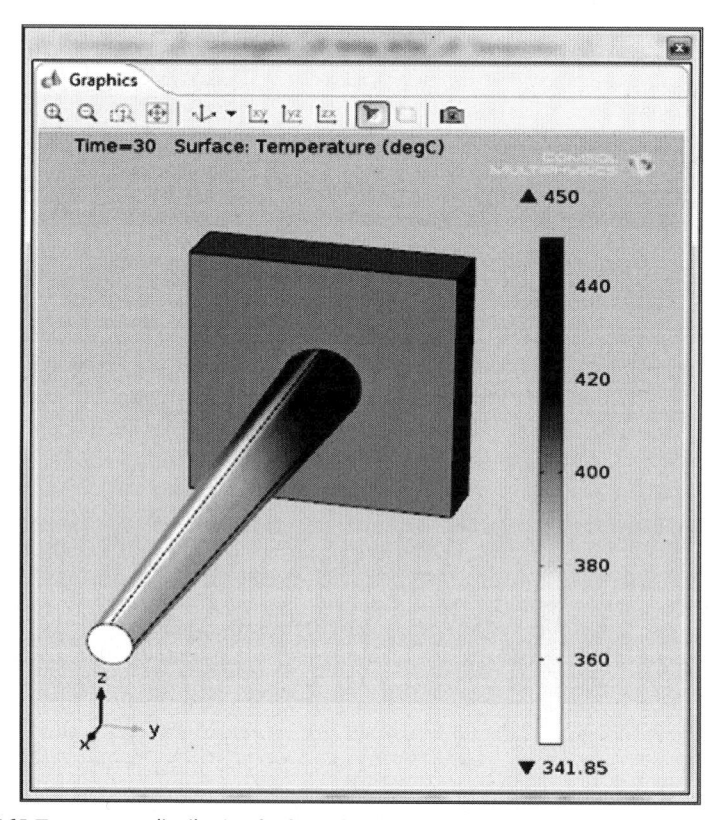

FIGURE 4.165 Temperature distribution for fin and wall section at time step 30.

13. To plot the temperature along the center-line, right-click on the Results node in the Model Builder and select 1D Plot Group. A new node, 1D Plot Group 1 will appear in the model tree. Right-click on this node and select Line Graph. In the Line Graph window, change Unit to **degC.** Click on the Define Cut Line icon and select two points along the fin axis. Notice the tool for selecting first and second point. Click Plot . The variation of temperature at different time steps (according to time step selection 0 to 60 seconds) for points along the selected line will appear. As expected, at early times temperature variations are much larger than those at later times. Results are shown in Figure 4.166.

14. Another useful plot is the variation of temperature versus time at a fixed point. Right-click on the Results node in the Model Builder and select 1D Plot Group. A new node, 1D Plot Group 2 will appear in the model tree. Right-click on this node and select Point Graph. In the Point Graph window, change Unit to **degC.** Select two points from the Graphics window, one at the fin base (e.g., point 9) and another one at the fin tip (e.g.,

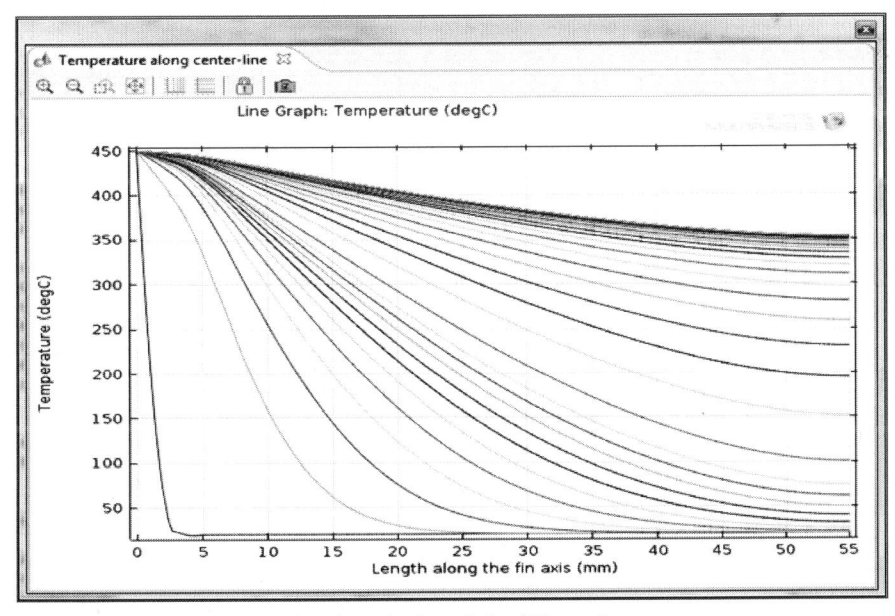

FIGURE 4.166 Temperature variation along the fins axis for different time steps.

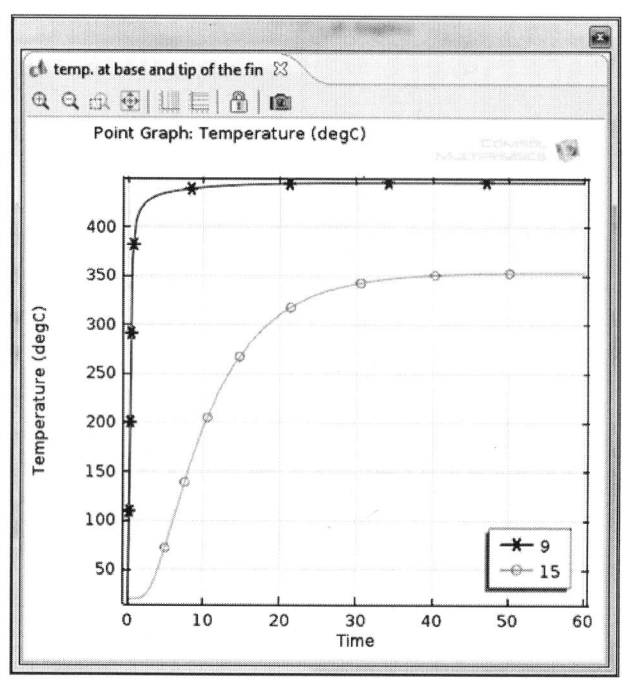

FIGURE 4.167 Temperature variations at the fin base (x) and its tip (o) versus time.

point 15) and add them to the Selection list using ✚. Expand Legends and check the box for Show legends. Click Plot. Results are shown in Figure 4.167.

Example 4.17: Heat Conduction Through a Multilayer Wall with Contact Resistance

In this example, we model heat conduction through a multilayer wall while considering thermal resistance between two of its layers. Contact resistance exists between dissimilar layers at their interface. The physics of contact resistance involves dimensions at micro scale (or even nano scale) level in order to capture the surface roughness of the two adjacent corresponding layers. For thermal engineering calculations, we consider an average (or statistically averaged) thermal resistance to represent the thermal contact resistance effects. However, the question is, "What is the thickness of this equivalent layer?" As mentioned above, this layer is very thin as compared to other actual corresponding layers' thicknesses. Therefore, if we create a very thin layer to represent the contact resistance it requires very small mesh size to build finite elements inside this layer, and therefore increases the number of elements and also creates unnecessary difficulties for the computations and numerical convergence of a model. One possible solution to this problem is to use conservation of energy (first law of thermodynamics) to conserve the heat flux going through the contact boundary. Therefore, the temperature will behave like a step function or will at least vary with very steep slope across the fictitious layer or the boundary. In COMSOL this tool is available as a type of boundary condition called Thin Thermally Resistive Layer (see COMSOL manual for details). The data required is an estimate of the layer thickness and optional thermal conductivities. If data is available, then users can enter it, otherwise the material properties of the adjacent layers will be used for calculations. Having this tool makes the calculations and inclusion of thermal resistance very simple as well as effective. However, users should note that realistic results require good data or estimates for the contact resistance interface layer and corresponding layers material properties.

In this example, we build a multilayer flat wall that includes a thin layer to represent contact resistance between two adjacent layers and calculate heat conduction and temperature profile through the wall section.

Solution

1. Launch COMSOL and open a new file. Save the file as Example 4.17.
2. In the Model Wizard window, select 2D and click Next.
3. From the Add Physics list, expand Heat Transfer and select Heat Transfer in Solids (*ht*). Add it to the Selected physics list by right-clicking and choosing Add Selected. Then click Next.
4. From the Select Study Type list, click on Stationary and then click on Finish.
5. In the Geometry window, select **mm** from the list for Length unit.
6. Draw a rectangle with dimensions Width 100 and Height 400, and set the Corner at x = y = 0. Rename the rectangle Brick. Similarly, draw four more rectangles with Widths 70, 30, 100, and 13, attached to each other with equal Heights of 400. Rename these layers Rain Screen, Air, Concrete, and Gypsum, respectively, as shown in Figure 4.168.

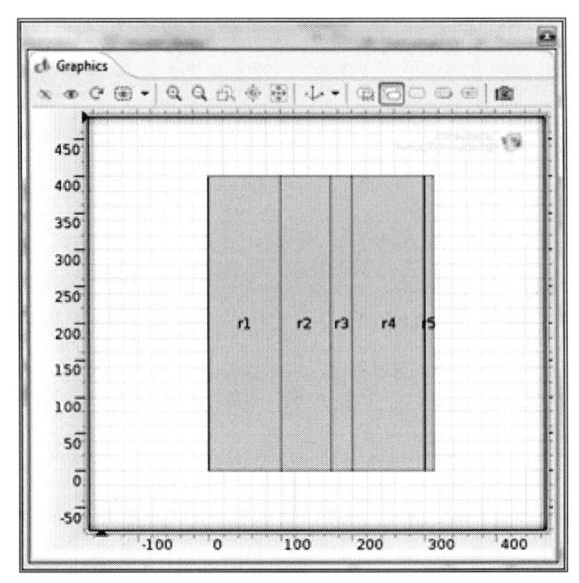

FIGURE 4.168 Window showing multilayer wall geometry.

7. Add materials to the wall model. Right-click on the Materials node in the Model Builder window. In the corresponding window, expand Built-In, select Brick, and add it to the model by clicking ✚. Similarly, for subsequent layers add Acrylic plastic, Air, and Concrete from the list. For the Gypsum layer, right-click on Materials node and select Material. In the corresponding window, enter the following data in the Material Contents section under Value: **k=0.276, Cp=1017, rho=711**. If needed, material property values can be changed for each layer by entering the desired value under the relevant Value column.

8. Define boundary conditions. Right-click on Heat Transfer in Solids (ht) and select Temperature from the list. In the Temperature window, add the left edge of the wall (boundary 1) to the Selection list. For Temperature, enter **-30[degC]**. Similarly, create another Temperature boundary condition and select and add the right edge of the wall (boundary 16) to its corresponding Selection list. For its Temperature, enter **20[degC]**.

9. Next we define the interface between the Brick and Rain screen layers as a thin layer with contact resistance. Right-click on Heat Transfer in Solids (ht) and select Thin Thermally Resistive Layer. In the corresponding window, add boundary 4 to the Selection. Locate the Thin Thermally Resistive Layer and enter the data as shown in Figure 4.169.

FIGURE 4.169 Contact resistance parameter setup.

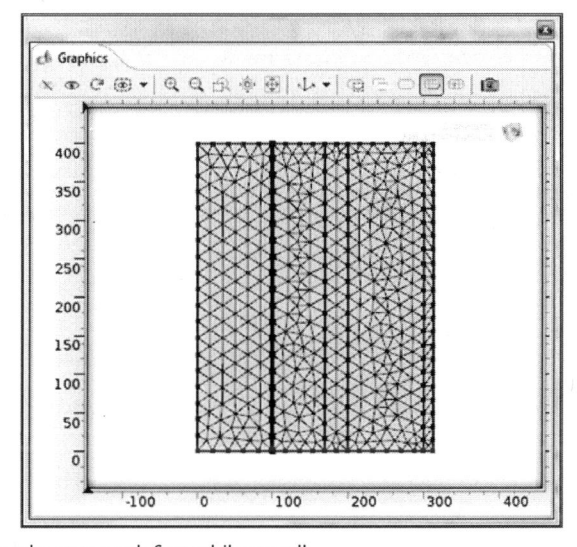

FIGURE 4.170 Finite elements mesh for multilayer wall.

10. To build a mesh, click on the Mesh 1 node in the Model Builder window. In the Mesh window, select Fine from the list for Element size. Click Build All. See Figure 4.170.

11. To run the model, right-click on Study 1 node in the Model Builder window and select Compute. Wait for computations to finish.

12. Default results for surface temperature and Isothermal contours appear in the model tree. Expand the Temperature (ht) node and click on Surface 1. In the Surface window, change Unit to **[degC]** and click Plot. Similarly, expand the Isothermal Contours (ht) node and click on Contour 1. In the Contour window, change Unit to **[degC]** and click Plot. Results are shown in Figure 4.171.

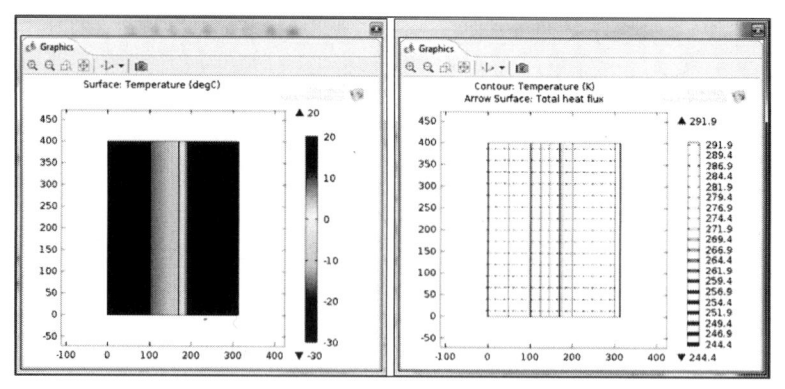

FIGURE 4.171 Windows showing results for temperature and heat flux through multilayer wall.

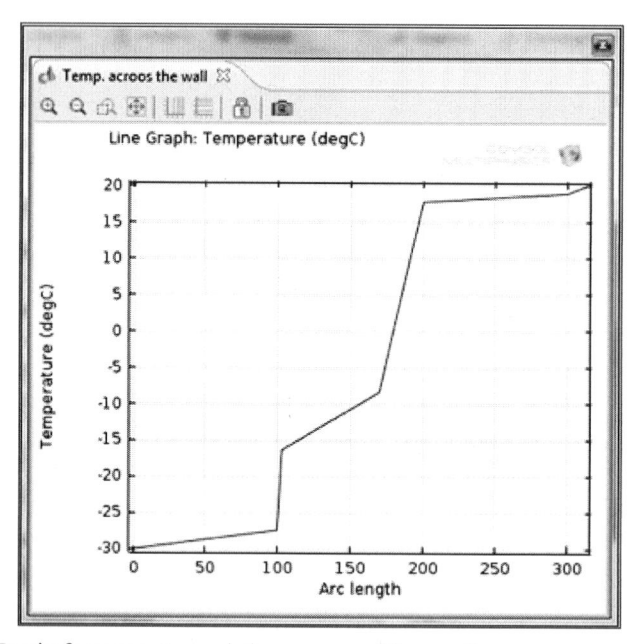

FIGURE 4.172 Results for temperature variations across multilayer wall.

13. It would be useful to make a graph of temperature variations across the wall. Right-click on the Results node in the Model Window and select 1D Plot Group. A new node 1D Plot Group 1 will appear in the model tree; right-click on it and select Line Graph. In the Line Graph window, change Unit to **[degC]** and click Plot. In this graph, as shown in Figure 4.172, the effect of contact resistance at interface located at 100 mm is clearly shown, which resulted in a sharp change in temperature.

SECTION 4.4: MODELING OF ELECTRICAL CIRCUITS

In this section, we use COMSOL modules and 0D features to model some examples of basic electrical circuits. In engineering, we encounter modeling problems that do not have any space or dimension dependency, such as chemical reactions, electric circuits, and Optimization. The 0D feature in COMSOL is used to model such problems.

Models discussed and built in this section include dynamic analysis and parametric study for typical RC and RLC electrical circuits. The main objective is to provide, for users and readers, some solved examples that can be used directly or lead to further solutions of similar or more complex structures using COMSOL. It is assumed that readers are familiar with relevant engineering principles and governing equations (see Reference 4.9). In each example, we provide brief explanations of physics involved along with governing equations and phenomena, as applicable. It is recommended that readers cover Chapters 1 through 3 before attempting examples in this section.

Example 4.18: Modeling an RC Electrical Circuit

In this example, we use a feature that is available in COMSOL 4.3, 0D (zero-dimension). We consider a simple resistor-capacitor electrical network, as shown in the Figure 4.173. We would like to calculate the voltage and current as a function of time.

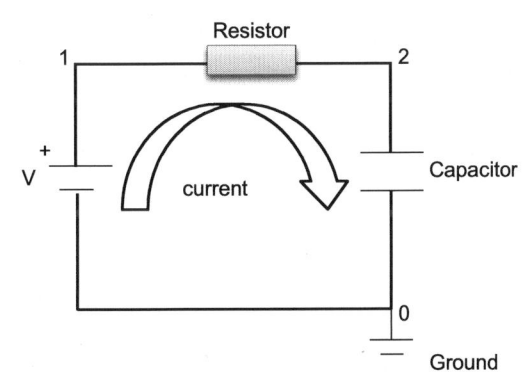

FIGURE 4.173 Sketch for the RC circuit and reference nodes used in the model.

Solution

1. Launch COMSOL and open a new file. Save the file as Example 4.18.
2. In the Model Wizard window, select 0D and click Next ⇨.
3. From the Add Physics list, expand AC/DC, select Electrical Circuit (*cir*), and add it to the Selected physics list by right-clicking and choosing Add Selected. Then click Next ⇨.
4. From the Select Study Type list, click on Time Dependent and then click on Finish.

For voltage calculation in an electrical network, a reference point is needed (similar to mechanical pressure). This point is referred to as Ground node. We set the Ground as node 0. For the purpose of identifying the components of the network as they connect to one another, we use data provided in Table 4.3, according to the sketch shown in Figure 4.173.

5. Expand Model 1 (*mod1*) node in the Model Builder window, then right-click on Electrical Circuit (*cir*) and select Voltage Source. In the corresponding window, enter data for Voltage and Node names as shown in Figure 4.174.
6. Again, right-click on Electrical Circuit (*cir*) and select Resistor. In the Resistor window, enter Node Names (1 and 2) and 5000[ohm] for Resistance.
7. To define the capacitor, right-click on Electrical Circuit (*cir*) and select Capacitor. In the Capacitor window, enter Node Names (2 and 0) and 10E-6 for Capacitance. Let Initial capacitor voltage be at the default value of zero.

The time step for this problem needs to be set. In an RC circuit, the time scale is the value of R x C (resistance times capacitance). Based on

Component	Start node	End node
Battery (voltage source)	0	1
Resistor	1	2
Capacitor	2	0

TABLE 4.3 Data for building the RC circuit nodes.

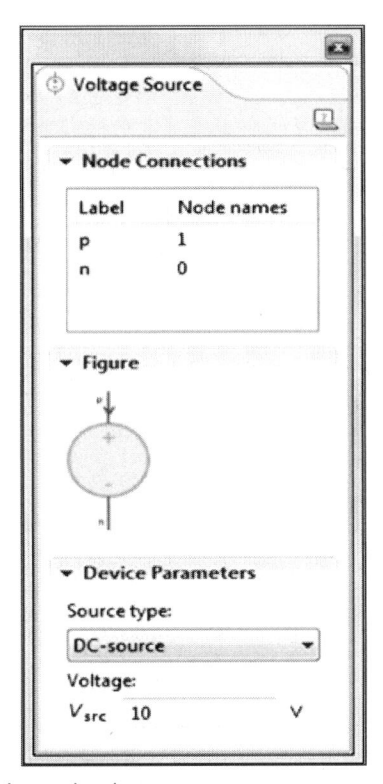

FIGURE 4.174 Voltage source data and nodes.

data provided, we have 0.05 seconds. Therefore, we run the model with a time step of 0.001 sec. and up to about five times the time scale, or 0.25 sec.

8. Expand the Study 1 node in the Model Builder window and click on Step 1: Time Dependent. In the corresponding window, enter **range(0, 0.001, 0.25)**. Right-click on Study 1 and select Compute ≣.

9. Right-click the Results node in the Model Builder and select 1D Plot Group. In the corresponding window, expand the Legend section and choose Lower right for Position. Right-click on 1D plot Group 1 and select Global. In the Global window under y-Axis Data section, click on ✚ and select Electrical Circuit>Voltage across device C1. Click Plot. The results, as shown in Figure 4.175, show the variation of voltage versus time. Graphs for voltage and currents for other circuit components can be plotted by choosing the corresponding variables for the list.

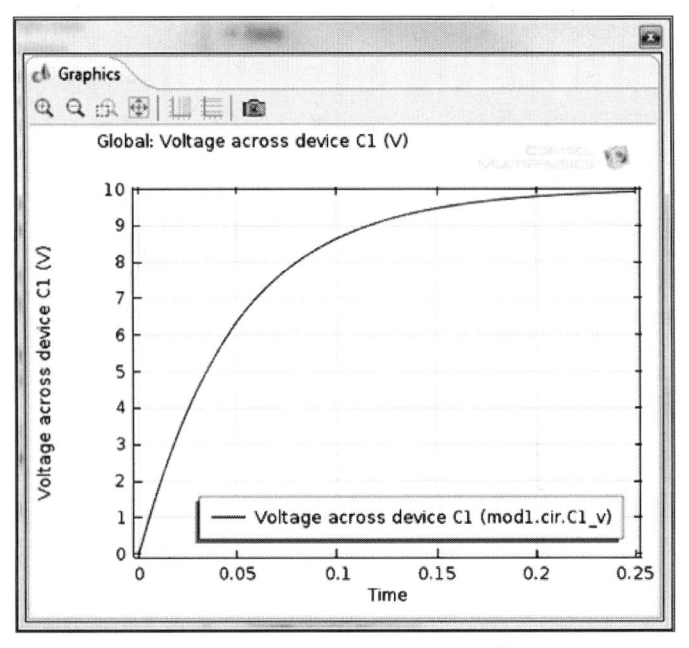

FIGURE 4.175 Results for voltage variation across capacitance device-C1.

Example 4.19: Modeling an RLC Electrical Circuit

An electrical circuit with resistor, capacitor, and inductor will be modeled in this example, using the 0D feature available in COMSOL. An RLC circuit behaves in oscillation like a mass-spring-damper mechanical system. To be exact, mass m is analogous to inductance L, spring stiffness k to inverse of capacitance C, damping coefficient γ to resistance R when we write the governing differential equation for electric current $i(t)$, which is analogous to mass displacement $x(t)$. A similar analogy can be drawn for voltage $v(t)$, as well. Corresponding ODEs are as follows:

$$m\frac{d^2x}{dt^2} + \gamma\frac{dx}{dt} + kx = f(t)$$

$$L\frac{d^2i}{dt^2} + R\frac{di}{dt} + \frac{1}{C}i = g(t)$$

$$C\frac{d^2v}{dt^2} + \frac{1}{R}\frac{dv}{dt} + \frac{1}{L}v = h(t)$$

The damping term is defined as $\alpha = \dfrac{R}{2L}$ for series and $\alpha = \dfrac{1}{2RC}$ for parallel RCL circuits.

Natural fundamental frequency for an RCL circuit is defined as $\omega_0 = (LC)^{-0.5}$. When $D = \alpha^2 - \omega_0^2$ is positive, the circuit is over damped; when equal to zero, the circuit is critically damped; and when negative, the circuit is under damped.

For this example, we consider a simple resistor-capacitor-inductor electrical network. We would like to calculate the voltage and current as a function of time.

The circuit sketch for this model is shown in Figure 4.176.

Solution

1. Launch COMSOL and open a new file. Save the file as Example 4.19.
2. In the Model Wizard window, select 0D and click Next ⇨.
3. From the Add Physics list, expand AC/DC, select Electrical Circuit (*cir*) and add it to the Selected physics list by right-clicking and choosing Add Selected. Then click Next ⇨.
4. From the Select Study Type list, click on Time Dependent and then click on Finish.

For calculating voltage in an electrical network, a reference point is needed (similar to mechanical pressure). This point is referred to as Ground

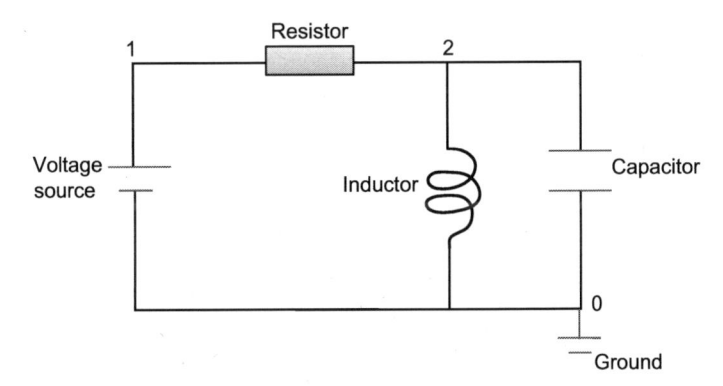

FIGURE 4.176 Sketch for the RCL circuit and reference nodes used in the model.

node. We set the Ground as node 0. For the purpose of identifying the components of the network as they connect to one another, we use data provided in Table 4.4, according to the circuit sketch.

5. Expand the Model 1 (*mod1*) node in the Model Builder window and right-click on Electrical Circuit (cir) and select Voltage Source. In the corresponding window, enter the data for Voltage and Node names as shown in Figure 4.177.

6. Again, right-click on Electrical Circuit (*cir*) and select Resistor. In the Resistor window, enter Node Names (1 and 2) and 3000[ohm] for Resistance.

7. To define the capacitor, right-click on Electrical Circuit (*cir*) and select Capacitor. In the Capacitor window, enter Node Names (2 and 0) and 1E-6 for Capacitance. Set Initial capacitor voltage at the default value of zero.

8. To define the inductor, right-click on Electrical Circuit (*cir*) and select Inductor. In the Inductor window, enter Node Names (2 and 0) and 100[mH] for Inductance. Set Initial inductor current at the default value of zero.

The time step for this problem needs to be set. In an RLC circuit, there are two parameters that define its time scale. The resonant frequency of the circuit is ω_0; for this circuit the value is about 3163 rad/s^{-1}. Therefore time scale is about 2E-3 seconds (inverse of $3163/2\pi$). Another time scale is set for the damping coefficient α. For this circuit the value is about 166.7 s^{-1} (=1/2RC, since capacitor and inductor are parallel). Therefore the time scale is about 6E-3 seconds. We choose the smaller value for time step equal to 0.002 s. Therefore we run the model with a time step of 0.0001 sec. and up to more than eight times the larger time scale or 0.05 sec.

Component	Start node	End node
Battery (voltage source)	0	1
Resistor	1	2
Inductor	2	0
Capacitor	2	0

TABLE 4.4 Reverence nodes used for the RCL circuit model.

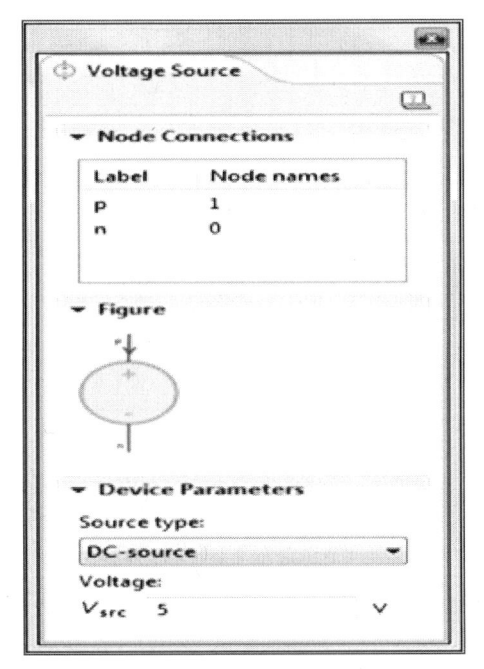

FIGURE 4.177 Voltage source data and nodes.

9. Expand the Study 1 node in the Model Builder window and click on Step 1: Time Dependent. In the corresponding window, enter **range(0, 1e-4, 0.05)**. Right-click on Study 1 and select Compute ≡ .

10. Right-click the Results node in the Model Builder and select 1D Plot Group. In the corresponding window, expand the Legend section and choose Upper right for Position. Right-click on 1D plot Group 1 and select Global. In the Global window under y- Axis Data section, click on ✛ and select Electrical Circuit>Voltage across device R1. Click Plot . The results show the variation of voltage versus time. Graphs for voltage and currents across other circuit components can be plotted by choosing the corresponding variables for the list. Results are shown in Figure 4.178.

We perform a parametric analysis for studying the effect of capacitance on the voltage output of the circuit.

11. Right-click on the Global Definitions node in the Model Builder window and select Parameters. In the Parameters window, enter **10[nF]**.

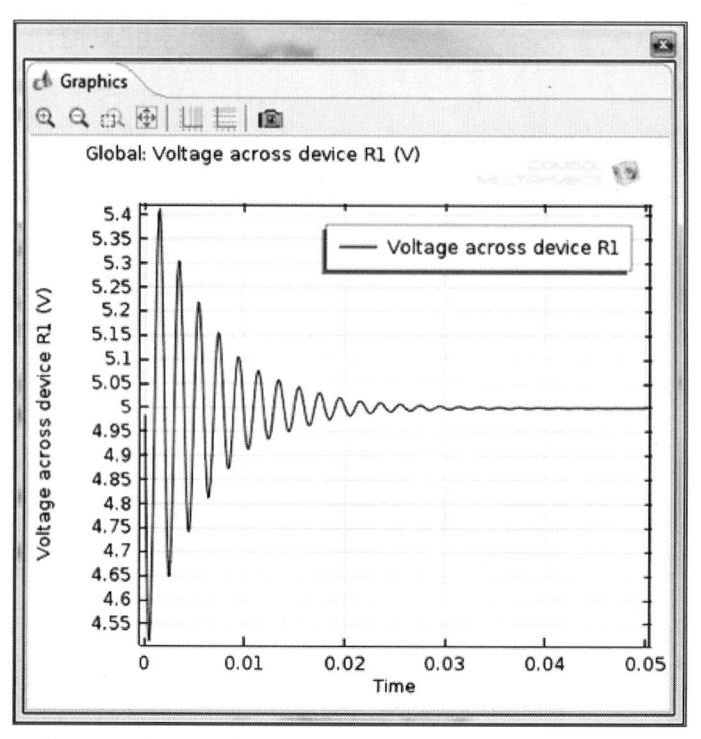

FIGURE 4.178 Voltage variation versus time across resistor device-R1.

12. Right-click on the Study 1 node and select Parametric Sweep. In the corresponding window, click on the plus icon and select C from the list under Parameter names. Enter **10[nF], 100[nF], 1000[nF]** for Parameter value list.

13. To run the model, right-click on the Study 1 node and select Compute ≡. Wait for computations to finish.

14. To show the results, right-click the Results node in the Model Builder and select 1D Plot Group. In the corresponding window, expand the Legend section and choose Lower right for Position. Right-click on the 1D plot Group 2 and select Global. In the Global window under Data, select Solution 2 from the list for Data set. Under y- Axis Data section, click on ✚ and select Electrical Circuit>Voltage across device R1. Click Plot. The results show the variation of voltage versus time. Graphs for voltage and currents across other circuit components can be plotted by choosing the corresponding variables for the list. Results are shown in Figure 4.179.

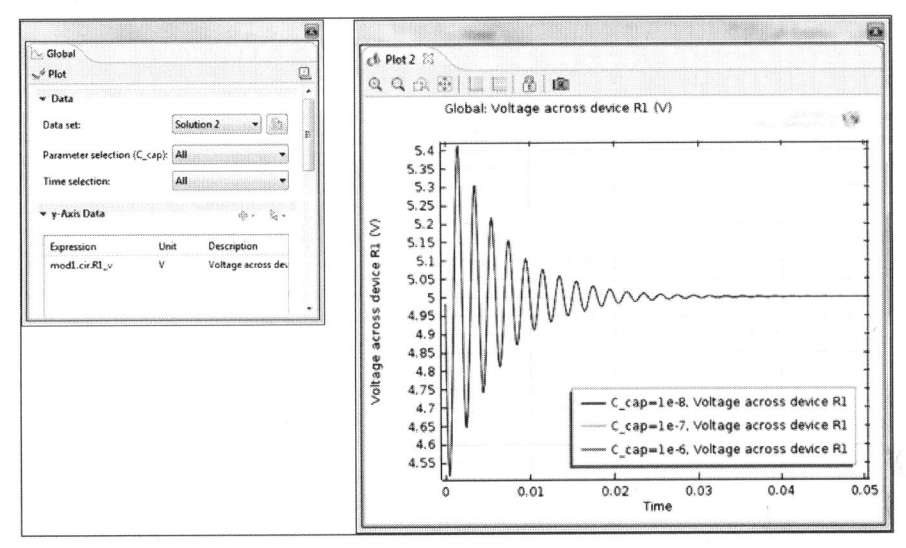

FIGURE 4.179 Windows showing voltage variation versus time across resistor device-R1 for several values of capacitances.

SECTION 4.5: MODELING COMPLEX AND MULTIPHYSICS PROBLEMS

In previous sections, we used COMSOL features to model problem examples that involved several different physical phenomena but mainly with one phenomenon dominating. Modern engineering involves problems that require solutions of multiphysics phenomena (see Reference 4.10). A multiphysics problem is defined as one that has two or more competing physics happening simultaneously. For example, when we have a fluid flowing over a relatively hot plate, heat is transferred to it and the temperature variations in the plate may change its thermal conductivity as well as that of fluid. This interactive effect should be accounted for in stress analysis of the plate and flow velocity and pressure calculations. Another example may be a cantilever beam in a fluid stream. The deflection of the beam, as a result of its stiffness, will affect the fluid's velocity and pressure and vice versa. Analyzing and modeling multiphysics problems can be done using COMSOL, which has many features and modules enabling users to "simply" add physics to a problem at hand.

In this section, we also solve complex problems for which users may have the relevant mathematical models (like ODEs and PDEs) developed and would like to have the solutions using finite element method with features available in COMSOL, such as a multibody dynamics problem or multiphysics problem for which COMSOL modules may not yet be available. We demonstrate the applications of finite element method to solve such problems through some examples. For examples in this section, previous knowledge (or review) of differential equations (both ordinary and partial) and Lagrangian mechanics (see Reference 4.11) would help users to benefit fully from their obtained solutions.

Example 4.20: Stress Analysis for an Orthotropic Thin Plate

This example extends the model built in Example 4.1. We consider an orthotropic material (i.e., material properties different along x and y axes). The plate carries stationary mechanical loads. It is recommended that readers review or build the model described in Example 4.1 before studying this example.

The plate geometry (12 in. by 8 in.) is given as shown in the Figure 4.180. All other dimensions are as those given in Example 4.1. We would like to calculate the displacement and stress field under the applied loads. We consider a plane stress case, which means that the stress component along the z-axis (normal to the plane of the plate) is zero, but not necessarily the strains. Plate thickness is 0.4 in. and properties of the material (for example Kevlar-epoxy) are given in Table 4.5. The boundaries on the left side of the plate are constrained while the other plate sides are let free.

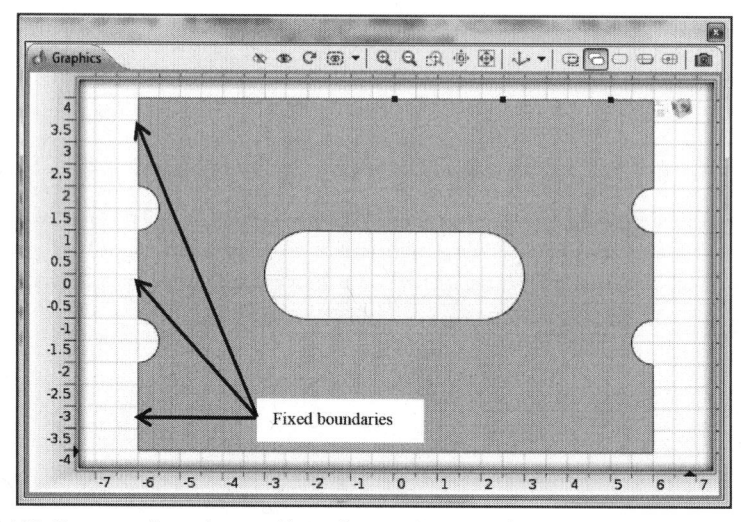

FIGURE 4.180 Geometry, dimensions, and boundary conditions.

Modulus of elasticity	Poisson's ratio	Shear modulus	Specific gravity
$E_x = 80$ GPa	$\nu_{xy} = 0.31$	$G_{xy} = 2.1$ GPa	1.44
$E_y = 5.5$ GPa	$\nu_{yx} =$ (calculated)		

TABLE 4.5 Material properties for orthotropic Kevlar-epxoy.

Before starting the model construction, we explain the material properties, since we are using an orthotropic material. From Table 4.5 we see that we have two moduli of elasticity. Obviously, this material is relatively stiffer in x-direction than in y-direction. We also have two Poisson's ratios. If we recall from isotropic materials, we usually have one modulus of elasticity and one Poisson's ratio. In order to explain the significance of orthotropic material properties, we use the stress-strain relationship resulting from general theory of elasticity. For our 2D plane stress problem at hand, the strains (ε's and γ) are related to stresses (σ's and τ) as:

$$\begin{Bmatrix} \varepsilon_x \\ \varepsilon_y \\ \gamma_{xy} \end{Bmatrix} = \begin{bmatrix} 1/E_x & -\nu_{yx}/E_y & 0 \\ -\nu_{xy}/E_x & 1/E_y & 0 \\ 0 & 0 & 1/G_{xy} \end{bmatrix} \begin{Bmatrix} \sigma_x \\ \sigma_y \\ \tau_{xy} \end{Bmatrix}$$

Due to symmetry we should have $-\nu_{yx}/E_y = -\nu_{xy}/E_x$, which is used in COMSOL for calculation of ν_{yx} $(= 0.31 \times 5.5 / 80)$ in this example. For further readings see Reference 4.12.

Solution

1. Open COMSOL and open the file Example 4.1.mph. If you have not built the model for Example 4.1, you can obtain it from the attached CD. Save the new file as Example 4.20 by clicking on File>Save As in the toolbar.

   In the Model Builder window, open the tree under Model 1 (*mod1*). Under Plate (*plate*), click on the Linear Elastic Material 1 node. In the Linear Elastic Material window, locate the Linear Elastic Material section and select Orthotropic from the list under Solid model. Under Young's modulus section, select User defined and enter the data for E as given in Table 4.5. Similarly, enter data for Poisson's ratio, Shear modulus, and density (1.44E3).

2. To run the model, right-click on the Study 1 node in the Model Builder window and select Compute ≡. The default result for von Mises stresses will appear in the Graphics window.

3. To manipulate the results, click on the Surface 1 node under the Stress Top (plate) node in the Model Builder window. This will open the Surface window. In this window, open the list under Unit in the Expression section and choose psi, then open the Title section and choose Manual from the Title type and enter **von Mises stress for Example 4.20 (psi).** Click (just once) on the (x-y) icon located on the Graphics window toolbar to set the view, then click on the Plot icon. Periodically, you may need to zoom in/zoom out to fit the results in the Graphics window. To show the displacement results, click on the Replace Expression icon (located on the right side of the Expression section) and choose Plate>Displacement>Total displacement (*plate.disp*). Change the Title type to Automatic and then click on the Plot icon. These results are shown in Figure 4.181 and Figure 4.182.

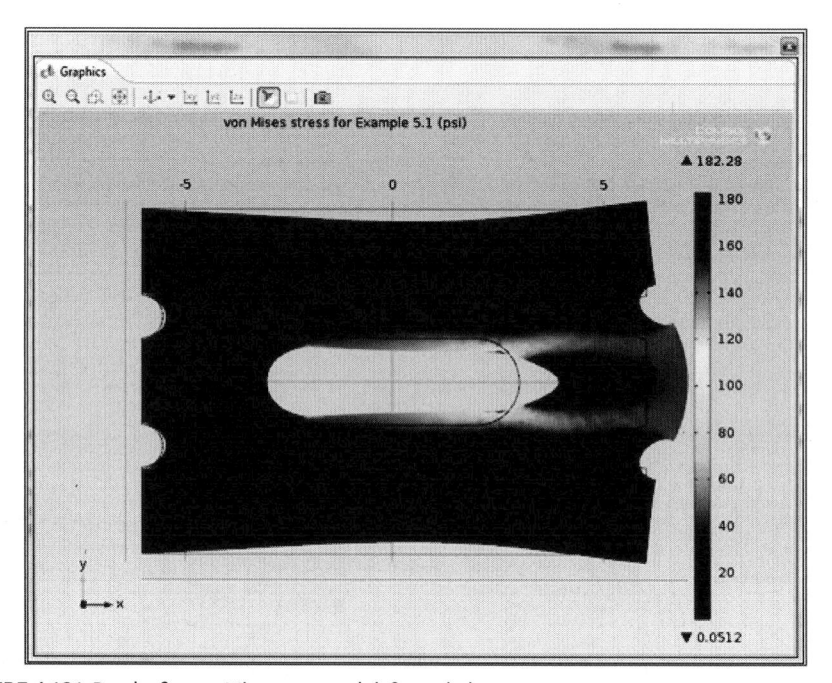

FIGURE 4.181 Results for von Mises stress and deformed plate.

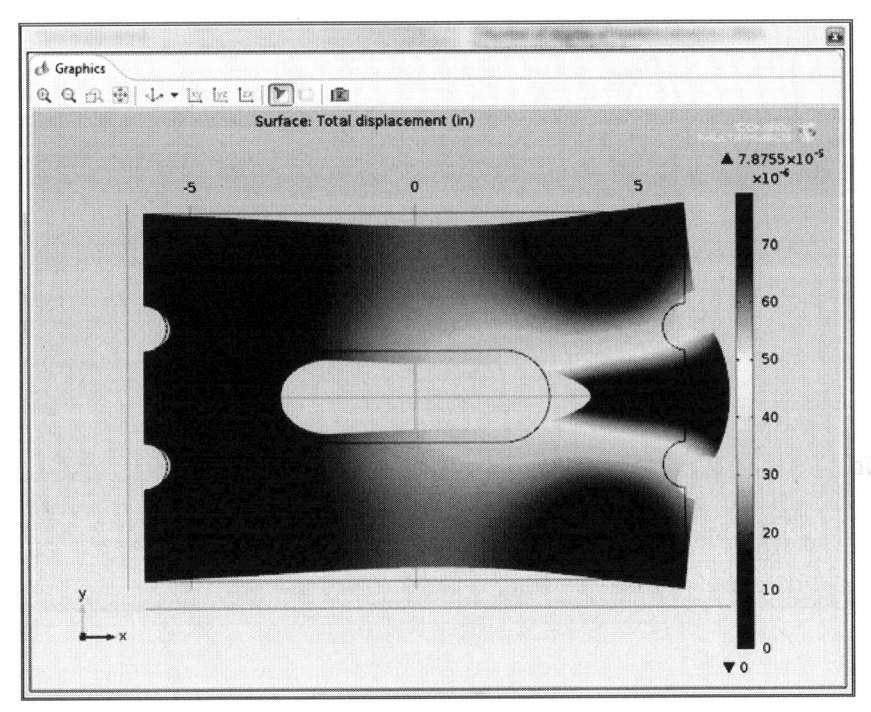

FIGURE 4.182 Results for total displacement for the orthotropic plate.

Example 4.21: Thermal Stress Analysis and Transient Response of a Bracket

In this example, we extend the model built in Example 4.3, the bracket. The first feature we consider is extending the model to calculate thermal stresses due to temperature variations at the boundaries. The second feature is modeling transient response of the bracket structure due to a transient load. For each modeling feature we use the basic file from the COMSOL model library (Structural_Mechanics_Module/Tutorial_Models/bracket_basic)[3] and add the corresponding model feature to it. Users can add these features directly to the file from Example 4.3, in Section 4.1.

---

[3] Model made using COMSOL Multiphysics® and is provided courtesy of COMSOL. COMSOL materials are provided "as is" without any representations or warranties of any kind including, but not limited to, any implied warranties of merchantability, fitness for a particular purpose, or noninfringement.

Solution

1. Launch COMSOL and open a new file.

2. Click on the Model Library tab. (If this window/tab is not visible, click on View in the main toolbar and select Model Library.) In the Model Library window, extend the Structural Mechanics Module, select **Tutorial Models>bracket basic**, and then click on Open Model to upload the model. Save this model as Example 4.21.

3. **Thermal stresses**: Right-click on the Model 1 (*mod1*) node in the Model Builder window and select Add Physics. The Model Wizard window will open; from the list select structural Mechanics>Thermal Stress (*ts*) and add it to the Selected physics list. Click on Next ⇨. From the Studies window, under Preset Studies select ▭ Stationary and click on Finish 🏁. Thermal Stress (*ts*) node will appear in the model tree, as shown in Figure 4.183.

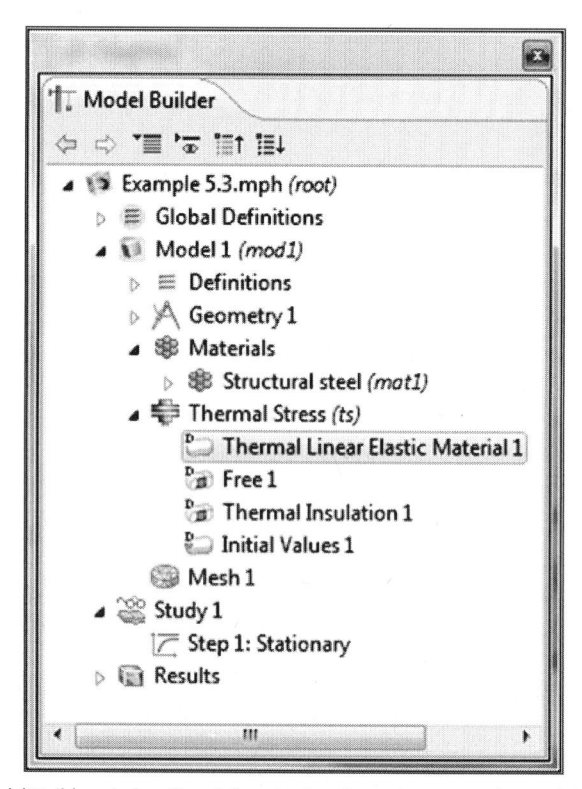

FIGURE 4.183 Model Builder window for adding physics, thermal stress, and materials.

4. Click on the Thermal Linear Elastic Materials 1 node. It shows the data entry tools for both elastic material data and thermal material data. COMSOL provides two methods for modeling thermal stresses. One approach requires building a model for structural analysis and calculating the thermal stresses due to thermal expansion, and finally coupling these two models. However, the Thermal Stress is a ready-to-use multiphysics interface that allows users to combine these features simultaneously, as used in this example. Right-click on the Thermal Stress (*ts*) node and select Solid Mechanics>Fixed Constraint from the Domain section (top group in the list). In the Fixed Constraint window, select Box 1 from the list under Domain Selection. This will constrain the bolts connectors to the bracket structure.

5. Thermal boundary conditions should be assigned as well. We set temperature at the bolt connectors at 20°C and those at the bracket arms where the circular holes are for pin holders at 100°C. The overall bracket is also cooled by convection or ambient air cooling. We set these three boundary conditions as follows. Right-click on Thermal Stress (*ts*) and select Heat Transfer>Convective Cooling. In the Convective Cooling window under the Heat Flux section, enter **10** for h and **15[degC]** for T_{ext}. To set the temperature boundaries, right-click on the Thermal Stress (*ts*) node and select Heat Transfer>Temperature. In the Temperature window, select boundaries that correspond to the bolts. For temperature T_o, enter **20[degC].** Similarly assign temperature boundary conditions for the pin holders on Box2. Right-click on Thermal Stress (*ts*) and select Heat Transfer>Temperature. In the Temperature window, select boundaries Box2 from the list under Selection and for temperature T_o enter **100[degC]**. See Figure 4.184.

6. Click on the Mesh node in the Model Builder. In the Mesh window, select Coarse for the Element size and click Build All.

7. Run the model by right-clicking on the Study node and select Compute. After the computation is finished, two 3D Plot Groups appear under the Results. Click on the first one from the top. The Graphics window shows the results for von Mises thermal stresses. Click on the second Plot Group to view the results for temperature in the Graphics window. For locating the Legend and changing the Units, click on the corresponding 3D Plot Group (or Surface) node to select the desired options. See Figure 4.185.

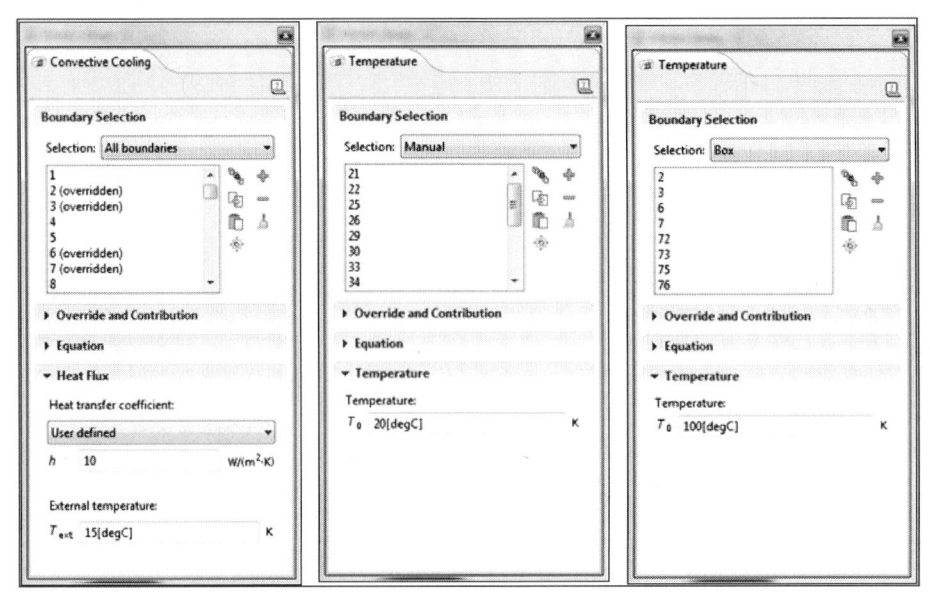

FIGURE 4.184 Windows showing boundary conditions setting for thermal stress calculations.

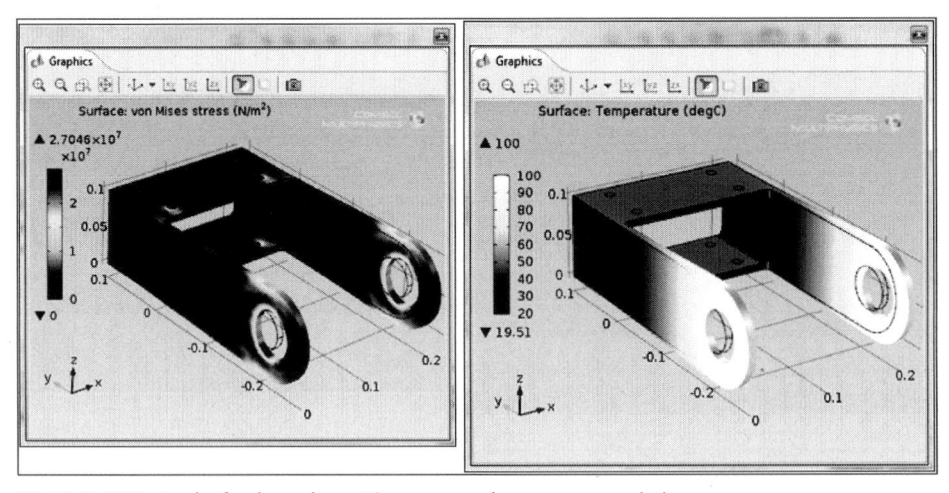

FIGURE 4.185 Results for thermal von Mises stress and temperature variations.

Transient response: Now we would like to model the bracket response to a set of given dynamic time-varying loads. Dynamic modeling of a structure requires definition for damping (the way a structure dissipates absorbed energy). In COMSOL, damping can be added to the model using Rayleigh damping or Loss Factor damping (see Reference 4.13). We use the Rayleigh damping method in this model. The Rayleigh method relates the damping coefficient c to the mass m (in general, mass matrix) and stiffness k (in general, stiffness matrix) using two parameters α_{dM} and β_{dK} as:

$$c = \alpha_{dM} m + \beta_{dK} k$$

8. Right-click on the Model (*mod1*) node in the Model Builder window and select Add Physics. In the Add Physics window, select Structural Mechanics>Solid Mechanics (*solid*) and click Next. In the Studies window, select Time Dependent and click on Finish.

9. Expand the Solid Mechanics (*solid*) node, right-click on the Linear Elastic Material node, and select Damping. In the Damping window, locate the Damping Settings section and choose Rayleigh damping from the list under Damping type. Enter **100** for α_{dM} and **3E-4** for β_{dK}, as shown in Figure 4.186.

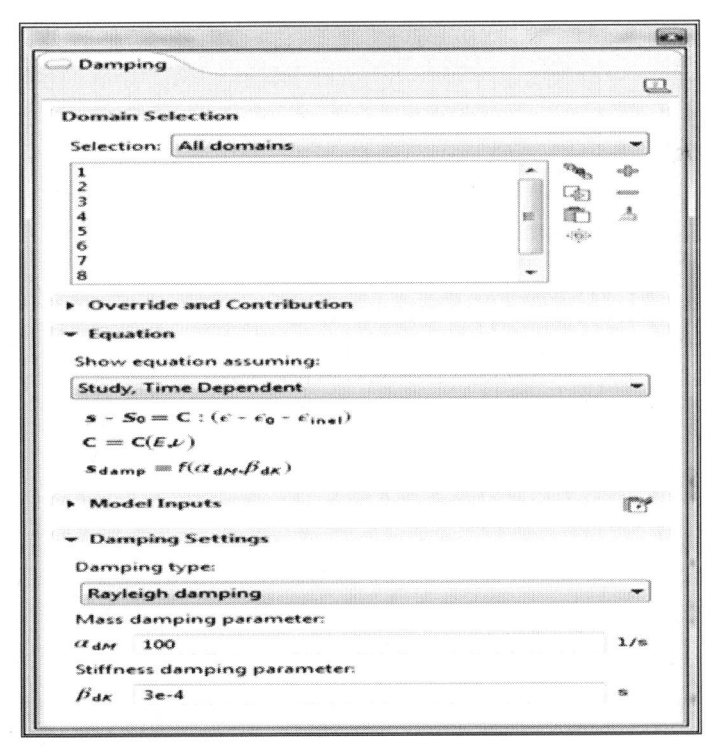

FIGURE 4.186 Rayleigh damping parameters setting.

10. Right-click on the Solid Mechanics (*solid*) node and select More>Fixed Constraint. In the Fixed Constraint window, select the bolt connectors (domains 2 to 9) and add them to the Selection list under Domain Selection. This will constrain the bolt connectors to the bracket structure.

11. Right-click on the Solid Mechanics (*solid*) node and select Rigid Connector. In the Rigid Connector window, select Box2 from the list to add them. This will represent the pin that connects the two arms of the structure, as shown in Figure 4.187.

12. The loads are applied on the pin connectors. The applied force in the x, y, and z directions are $F_x = 10^3 \sin(2\pi \cdot 200 \cdot t)$, $F_y = -10^5$, and $F_z = 10^4 \sin(2\pi \cdot 100 \cdot t)$, respectively. To enter these forces, right-click on the Rigid Connector node in the Model Builder and select Applied Force. In the Applied Force window, locate the Applied Force section and enter 1e3*sin(2*pi*t*200[1/s]) for x, −1e5 for y, and, 1e4*sin(2*pi*t*100[1/s]) for z.

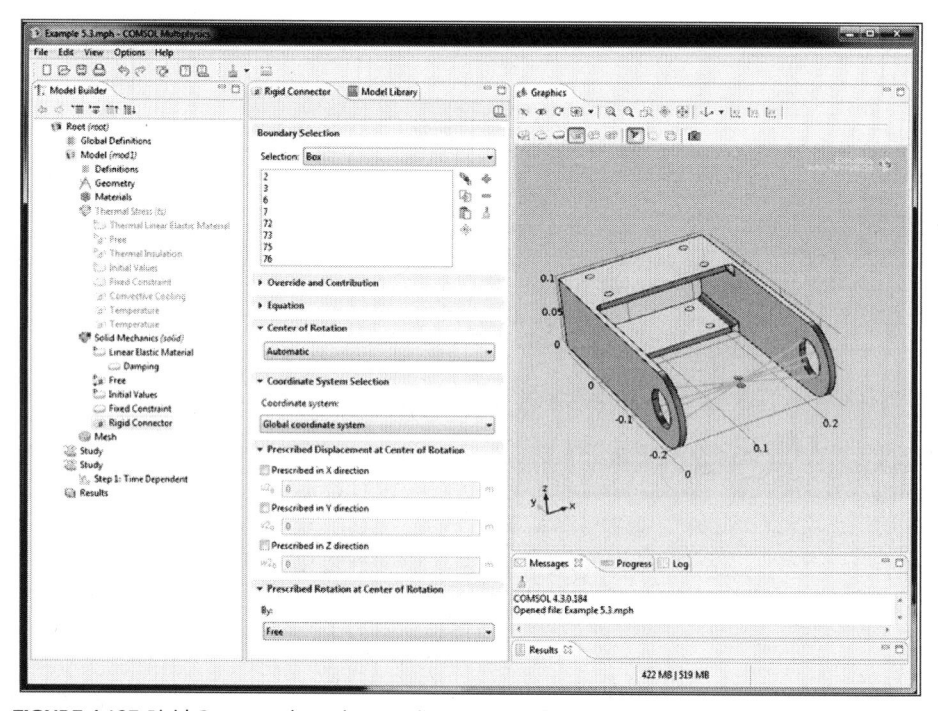

FIGURE 4.187 Rigid Connector boundary conditions setting for the bracket.

13. To disengage the thermal stress analysis, right-click on the Thermal Stress (*ts*) node and select Disable.

14. We set up a range of values for time variable *t*, for saving the results. Click on Study>Step 1: Time Dependent. In the Time Dependent window, enter **range(0, 0.25[ms], 15[ms])** in the space provided for Times under Study Settings. This will tell the solver to store the results for 15 ms at 0.25 ms intervals. Click the box in front of Relative tolerance and enter 0.001 in the space provided. This will ensure that the solver takes small time intervals during the transient solution. Also expand Results While Solving and check the box for Plot.

15. Click on the Show icon in the Model Builder toolbar and select Advanced Study Options. In the Model Builder window, right-click Study 2 and choose Show Default Solver. Expand Study>Solver Configurations>Solver>Dependent Variables. Click on **mod1.u2** and in the Field window locate the Scaling section and enter **1e-4** for Scale, as shown in Figure 4.188.

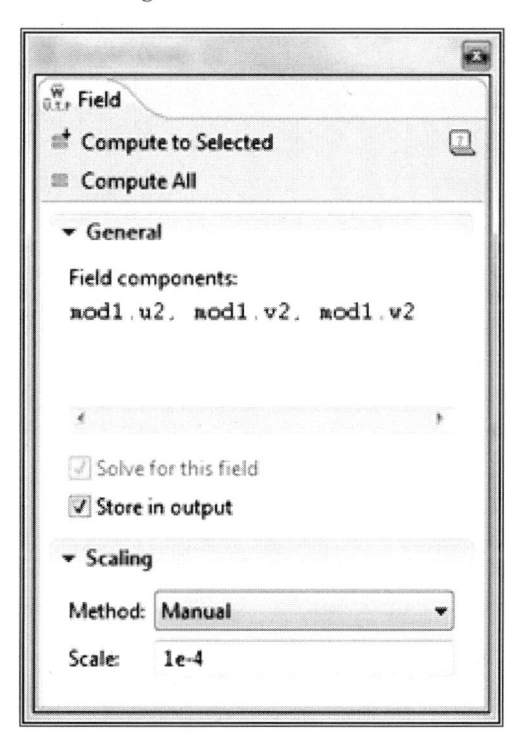

FIGURE 4.188 Solver parameters setting.

16. Right-click on Study and select Compute ≡. Wait until computations are done.

17. Click on Stress (*solid*) under Results node in the Model Builder window. In the corresponding window, locate Data section and select Solution 2 from the Data set: list and 0.015 from the Time list. Click Plot. The results show von Mises stress due to applied loads at time equal to 0.015 seconds, as shown in Figure 4.189. It would be useful to animate the displacements. Click on the Player icon in the toolbar. A movie of the displacements will be displayed in the Graphics window. Click on the Player 1 node in the Model tree to change the parameters of the animation.

18. To graph the displacements, right-click on Results and select 1D Plot Group. Click on 1D Plot Group 4 node; in the corresponding window locate Data section and select Solution 2 from the Data set: list. In the same window, expand Legend and select Upper left from the Position list. Right-click on the 1D Plot Group and select Global. Click on the Global1 node and in the corresponding window enter **solid.u_rig1, solid.v_rig1, soli.w_rig1** in the table under y-Axis Data section. Click Plot. Results are shown in Figure 4.190.

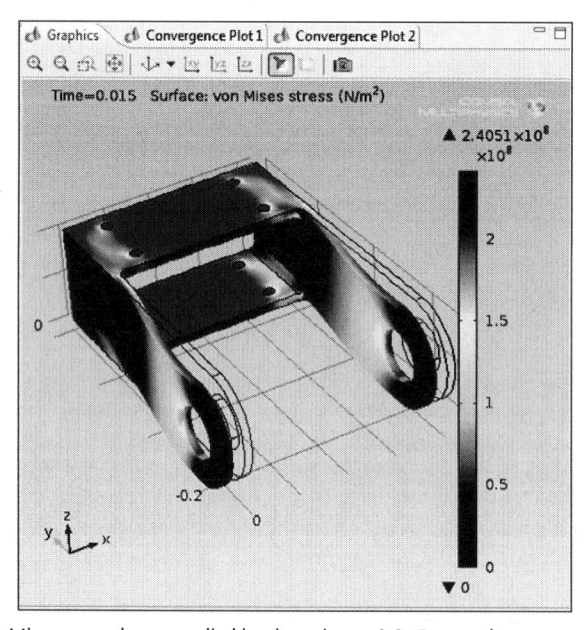

FIGURE 4.189 von Mises stress due to applied loads at time = 0.015 seconds.

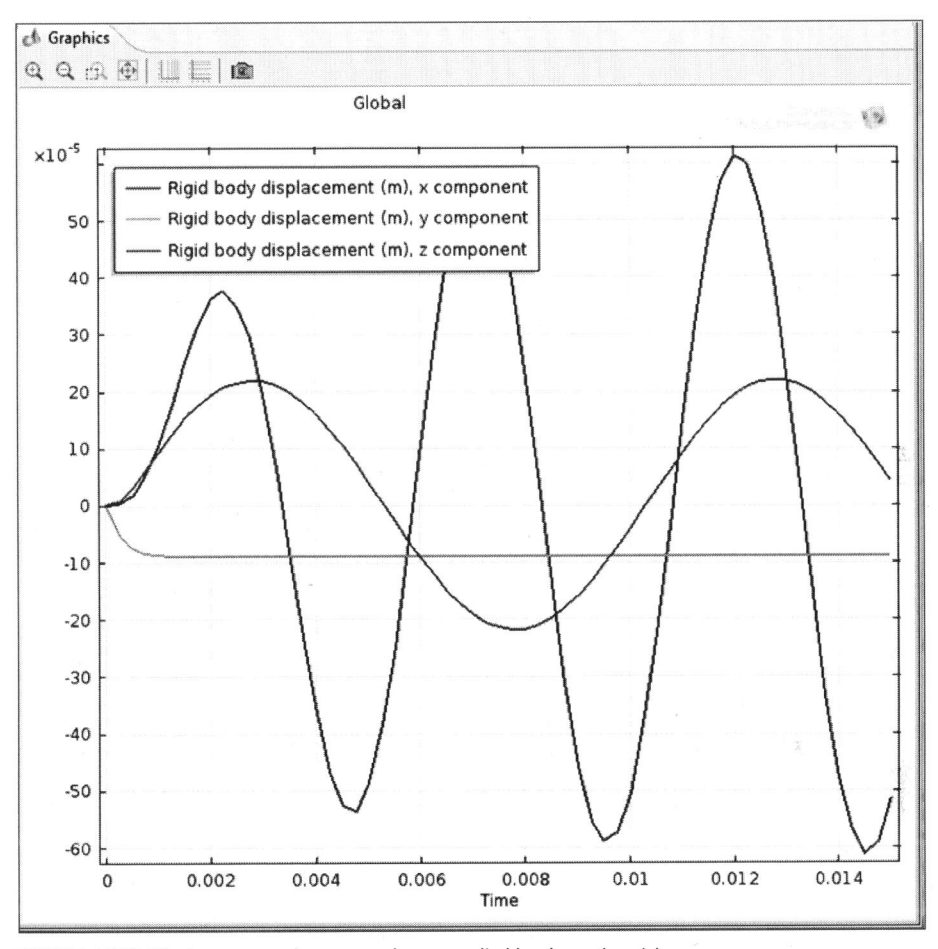

FIGURE 4.190 Displacement components due to applied loads vs. time (s).

Example 4.22: Static Fluid Mixer with Flexible Baffles

In many industrial machines and processing facilities, we may have cases in which a fluid flow interacts with solid structures or parts. We usually, as an approximation, consider the involved structure as a rigid body and calculate quantities like drag force on the structure. In many practical applications, the flexibility of the structure should be taken into account for such calculations. Therefore the deformed structure, as a result of exerted fluid forces, will change the fluid velocity and vice versa. This type of modeling is called *fluid-structure interaction*, abbreviated in COMSOL as *fsi*. The calculation of interaction of a solid with a fluid flow is, in principle,

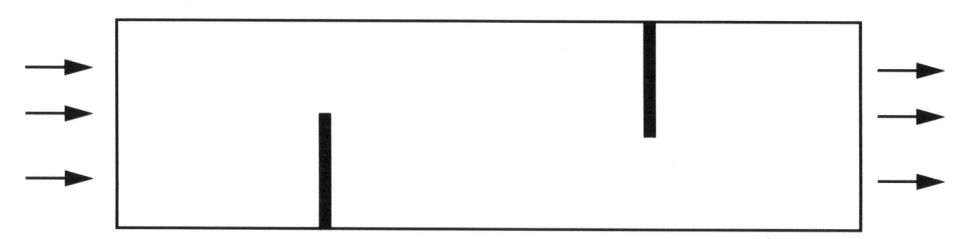

FIGURE 4.191 Sketch of the geometry for fluid mixed with baffles.

an iterative process and a very lengthy one. In this example, we model a 2D static mixer with two flexible baffles (beams) that are attached to the walls of the mixer and behave like cantilever beams. The deformations, von Mises stresses for the solid beams, and fluid velocity and pressure fields are calculated. The governing equations are equilibrium and Navier-Stokes equations. Refer to the COMSOL Manual for further readings.

A sketch of the problem geometry is shown in Figure 4.191. We build this geometry using CAD tools available in COMSOL.

Solution

1. Launch COMSOL and open a new file. Save it as Example 4.22.

2. Select 2D from the Model Wizard window and click. In the Add Physics window, expand the list for Fluid Flow and select Fluid-Structure Interaction (*fsi*). Then click on Next to go to the next window, Selected Study Type. In this window, click on Stationary node and click on the flag icon to Finish.

3. To build the geometry of the mixer, use the following steps:

    3.1 Click on the Geometry 1 node in the Model Builder window (if not already highlighted). In the Geometry window under Units, change the Length unit to **cm** from the list (open the list to see the selection options). Draw a box in the Graphics window by right-clicking on the Geometry 1 node and selecting Rectangle. The Rectangle window will open. In this window, enter 50 for Width and 15 for Height in the Size section, and 0 for x: and 0 for y: under Position. Make sure that Corner is selected for Base. Click on the Build Selected icon. The mixer box geometry will appear in the Graphics window.

3.2 Similarly, draw two more boxes for the beams/baffles. Right-click on the Geometry 1 node in the Model Builder window and select Rectangle. In the Rectangle window, enter 1 for Width and 8 for Height in the Size section, and 16 for x: and 0 for y: under Position. Make sure that Corner is selected for Base. Click on the Build Selected icon. Again, right-click on the Geometry 1 node in the Model Builder window and select Rectangle. In the Rectangle window, enter 1 for Width and 8 for Height in the Size section, and 32 for x: and 7 for y: under Position. Make sure that Corner is selected for Base. Click on the Build Selected icon. Finally, to smooth the tips of the beams just created, right-click on the Geometry 1 node in the Model Builder window and select Fillet. In the Fillet window, select the points on the tips of the beams/baffles and add them to the Selection window. Enter 0.5 for Radius, and then click on the Build Selected icon.

3.3 Rename the nodes under the Geometry 1 node as mixture, baffle 1, and baffle 2, as shown in Figure 4.192.

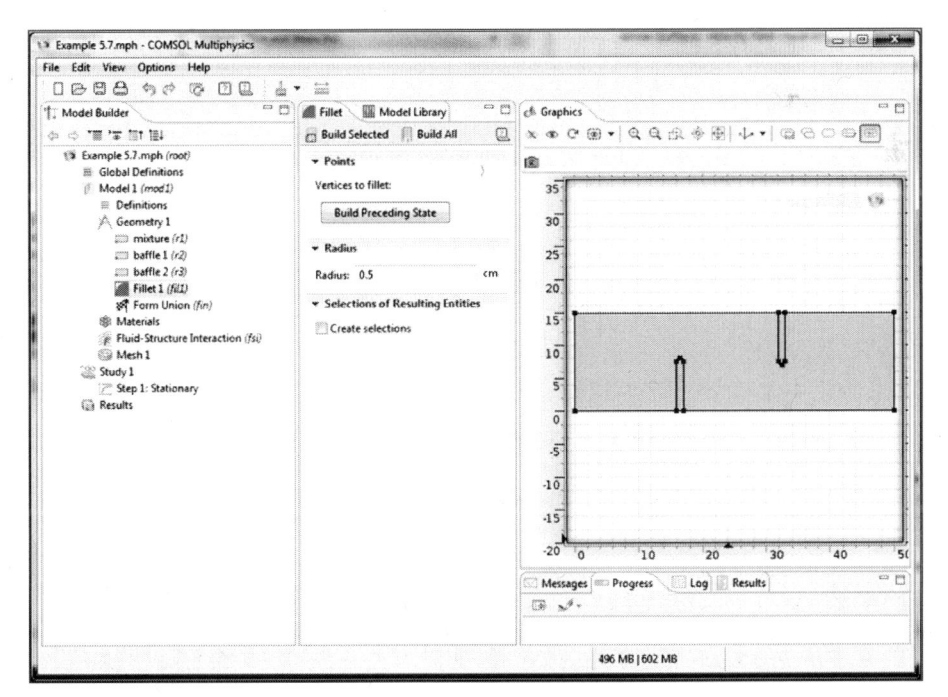

FIGURE 4.192 Building the mixer geometry.

4. To assign materials properties, we start with the baffles/beams. To define the baffles as solid media beams, expand the Fluid-Structure Interaction (*fsi*) node in the Model Builder and click on the Linear Elastic Material 1 node. From the Graphics window, select and add the domains (2, 3) for the baffles to the Selection list in Linear Elastic Material.

To assign the baffles material properties, right-click on the Materials node and select Material. In the Material window, make sure that only domains 2 and 3 are listed, as they represent the baffles/beams. Enter the following data for the beams properties: Young's modulus of elasticity (5.6e6), Poisson's ratio (0.4), and density (1e3). See Figure 4.193.

Similarly, assign fluid domain material properties. Right-click on the Materials node and select Material. In the Material window, make sure that only domain 1 is listed, as it represents the fluid flow domain. Enter 1E3 for fluid density and 1 for dynamic viscosity.

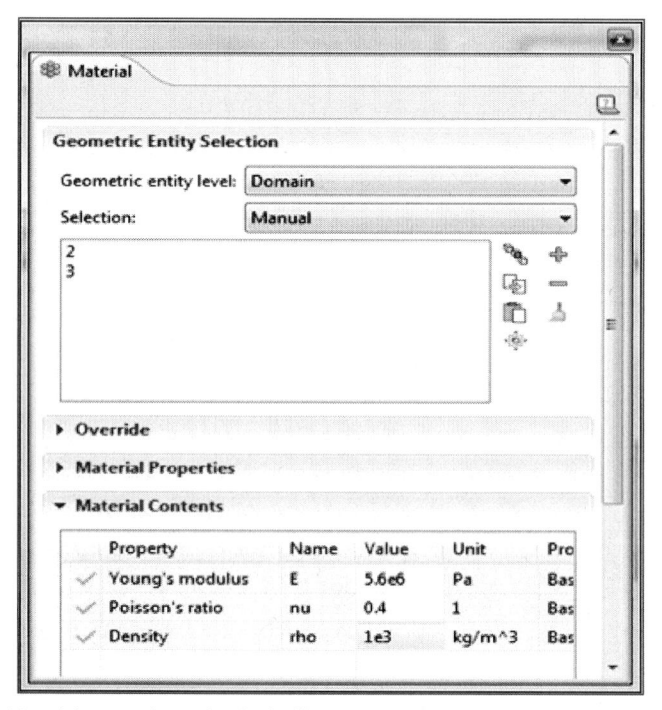

FIGURE 4.193 Material properties setting for baffles.

Finally, rename Material 1 to Material-baffle and Material 2 to Material-fluid.

5. To assign the boundary conditions, we first assign those related to the beams/baffles. Right-click on the Fluid-Structure Interaction node and select Solid Mechanics>Fixed Constraint. From the Graphics window, select the boundaries (5, 9) at the base of the beams where they are attached to the mixer wall, and add them to the Selection list in the Fixed Constraint window. This step is very crucial in finding a converged solution for this model, and in general fluid-structure interaction models. See Figure 4.194.

6. Assign the boundary conditions for the flow. Right-click on the Fluid-Structure Interaction node and select Laminar Flow>Inlet. In the Inlet window, add the boundary (1) to the Selection list. Select Laminar Flow from the list under Boundary Condition. Enter **Vin** for U_{av} and 10 for L_{entr}. Check the box for Constrain endpoints to zero. Similarly for outlet boundary condition, right-click on the Fluid-Structure Interaction node and select Laminar Flow>Outlet. In the Outlet window, add the boundary (12) to the Selection list. We leave the pressure as zero, for reference.

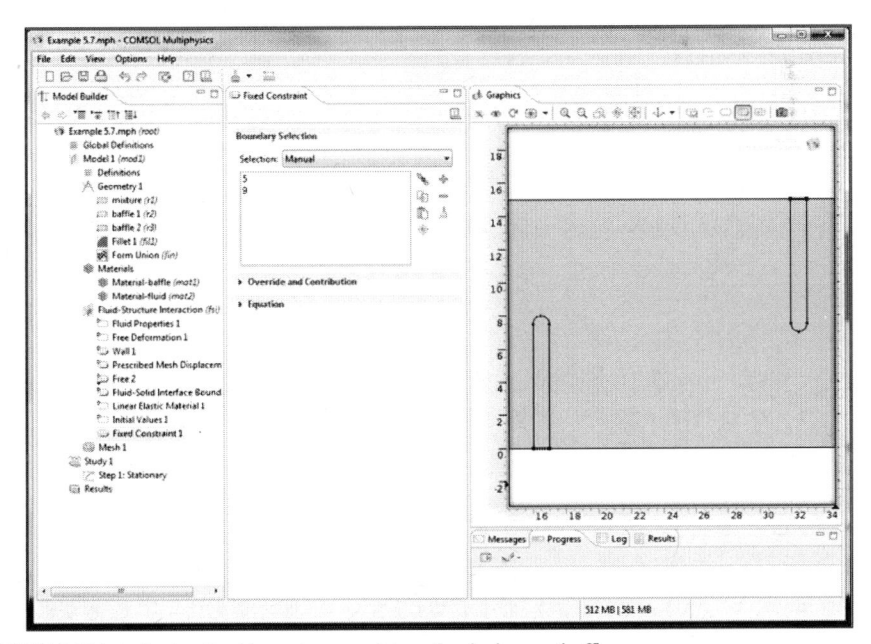

FIGURE 4.194 Assigning fixed boundary conditions for the beams/baffles.

7. To define the parameter Vin, the fluid velocity at the inlet, right-click on the Global Definitions node in the Model Builder window and select Parameters. In the Parameters window, enter **Vin** for Name and **0.2[m/s]** for Expressions. You may want to add Fluid Velocity under Description (optional).

At this stage, we have defined all the physics and boundary conditions for this model. Note that COMSOL will assign a moving mesh that moves according to the fluid flow and beams displacements. This feature is very useful for calculating the interactions and is done automatically. It is recommended that at this point users click on each of the nodes listed under the Fluid-Structure Interaction node in the Model Builder window and study their definitions.

8. We change the default discretization or the order of function/ polynomials for finite elements for fluid flow. We want to use second order for velocity and first order for pressure. This is defined as P2 + P1. To do this, click on the Show button, located in the Model Builder window toolbar, and select Discretization. Now click on the Fluid-Structure Interaction node and in the corresponding window locate the Discretization section and expand it. From the list under Discretization of fluids, select P2 + P1.

9. To build the mesh, click on the Mesh 1 node in the Model Builder window. In the Mesh window, change the Element size to Coarse and click Build All. 4170 elements are created. Zoom on the mesh to see the hybrid mesh, which consist of boundary-layer quadrilateral and main domain triangular elements. Also note that the beam/baffle solid domain is meshed, as shown in Figure 4.195.

10. The model is now ready to run. Right-click on the Study 1 node and select Compute. After the calculations are complete, the default results appear in the Graphics window, showing the fluid velocity and the von Mises stress in the beams/baffles for the assigned value of the inlet velocity 0.2 m/s.

11. It is useful to do a parametric analysis for this model based on the inlet velocity. Right-click on the Study 1 node and select Parametric Sweep. In the corresponding window, locate the Study settings section and click on the plus icon; the **Vin (Fluid velocity)** will appear in the table. This is the same as it was defined in the Global

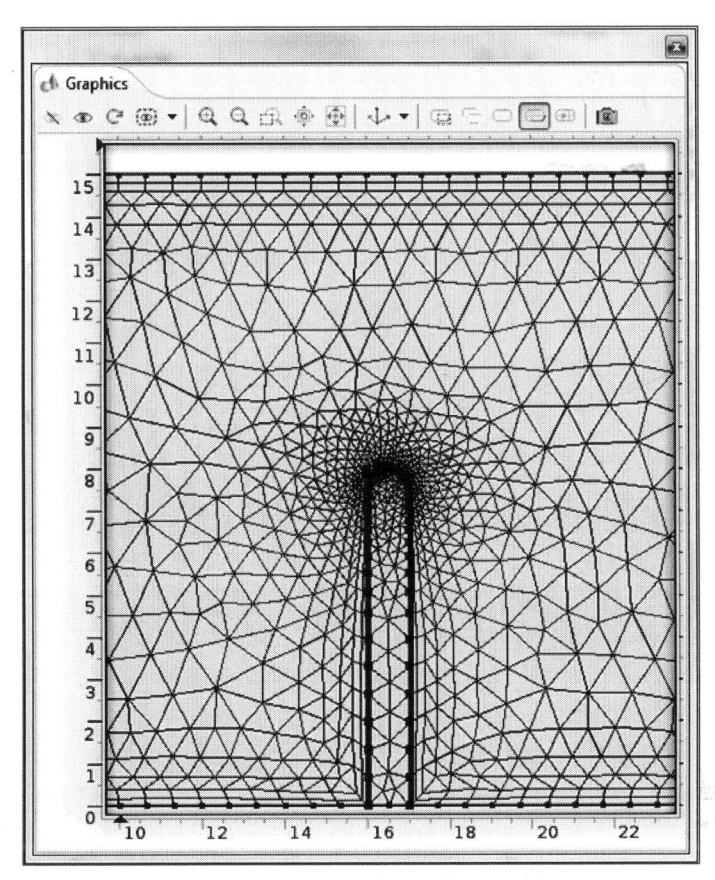

FIGURE 4.195 Built hybrid mesh, zoomed around the first baffle.

definitions>Parameters (see Step 7). In the Table under the Parameter value list, enter **0.2, 0.4, 0.6, 0.8.** The model will assign these values to the inlet velocity, in turn, and calculate the corresponding results. To see the results while being solved, check the Plot box under Output While Solving. This will slow down the solution process because it requires more computer resources.

12. Again, right-click on the Study 1 node and select Compute. The results will appear in the Graphics window. To see each set, click on the Flow and Stress node in the Model Builder window and in the corresponding window select the desired value for Parameter value (Vin), under the Data section.

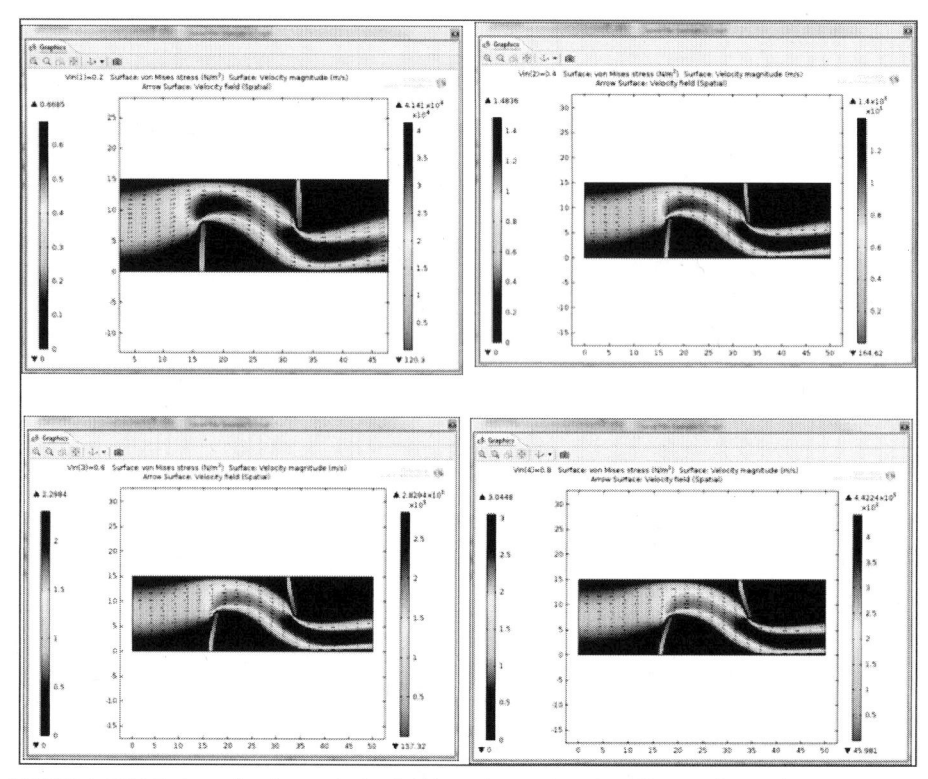

FIGURE 4.196 Windows showing results for fluid-beam interactions for different values of fluid inlet velocity.

13. The results for V_{in} = 0.2, 0.4, 0.6, and 0.8 m/s are shown in Figure 4.196. Note the progressive deflections of the beams/baffles as the fluid flow moves faster.

14. It would be useful to draw the displacements of the tips of the baffles/beams versus the inlet velocity of the fluid. Right-click on the Results node and select 1D Plot Group. A new node, 1D plot Group 3, will be created. Right-click on it and select Point Graph. In the corresponding window, select the points (5, 10) representing the tip points of the baffles and add them to the Selection list. In the y-axis Data section, enter **fsi.disp.** This variable is the total displacements as a result of beam deflection. Change the unit to **mm**, by selecting it from the Unit list. To draw the results, click on Plot . See Figure 4.197 and Figure 4.198.

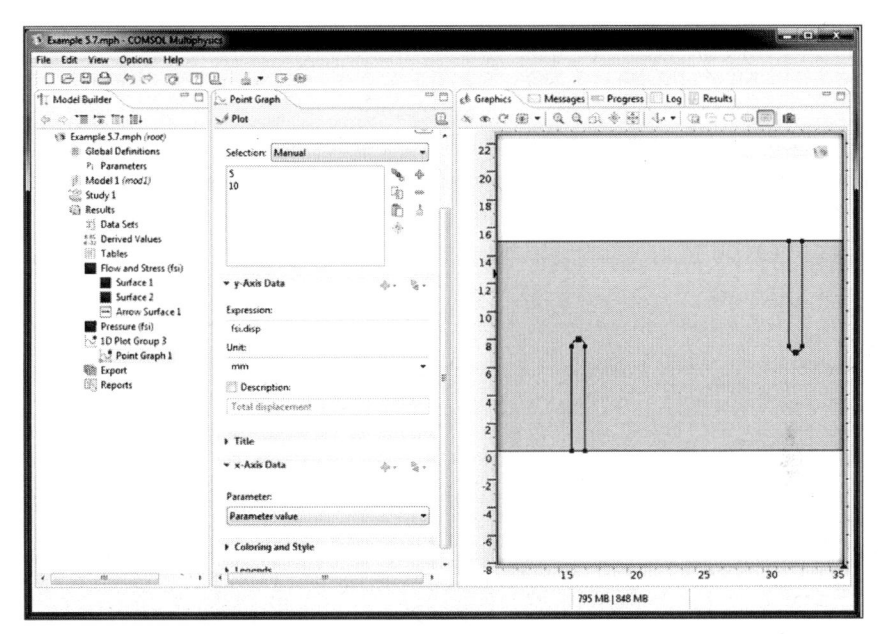

FIGURE 4.197 Building a graph for drawing tip displacements of the baffles.

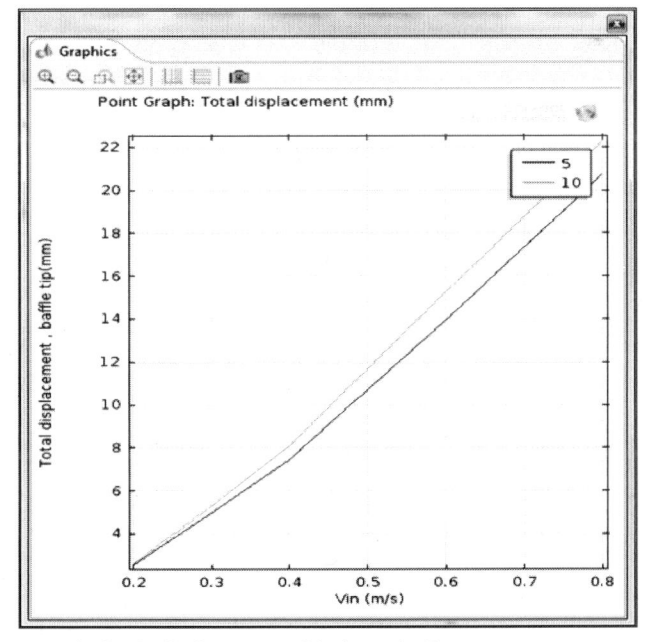

FIGURE 4.198 Results for tip displacements of the beam/baffles.

Example 4.23: Double Pendulum: Multibody Dynamics

In this example, we consider motion of a double pendulum. This is a classic problem in applied mechanics under multibody dynamics topic. As shown in Figure 4.199, the double pendulum system consists of two masses (m_1 and m_2) hanging by two strings with lengths L_1 and L_2. We assume that strings are massless, relatively. Vibration is in the x-y plane. We are mainly interested in finding the variations of angles θ_1 and θ_2.

The governing equations for motions of masses m_1 and m_2 are derived using Lagrangian mechanics (see Reference 4.11). The Lagrangian function is defined as $L = K - V$, where K is the kinetic energy and V potential energy of the system. For the double pendulum, we have

$$K = \frac{1}{2}m_1(\dot{x}_1^2 + \dot{y}_1^2) + \frac{1}{2}m_2(\dot{x}_2^2 + \dot{y}_2^2)$$

$$V = -m_1 g L_1 \cos(\theta_1) - m_2 g (L_1 \cos\theta_1 + L_2 \cos\theta_2)$$

where $x_1 = L_1 \sin\theta_1$, $x_2 = x_1 + L_2 \sin\theta_2$, $y_1 = -L_1 \cos\theta_1$, $y_2 = y_1 - L_2 \cos\theta_2$ with reference to the coordinates system shown in Figure 4.199. Dot-symbols are used for time derivatives of variables. By substituting K and V into L, after simplification we have:

$$L = \frac{1}{2}(m_1 + m_2)L_1^2 \dot{\theta}_1^2 + \frac{1}{2}m_2 L_2^2 \dot{\theta}_2^2 + m_2 L_1 L_2 \dot{\theta}_1 \dot{\theta}_2 \cos(\theta_1 - \theta_2)$$
$$+ (m_1 + m_2)g L_1 \cos\theta_1 + m_2 g L_2 \cos\theta_2$$

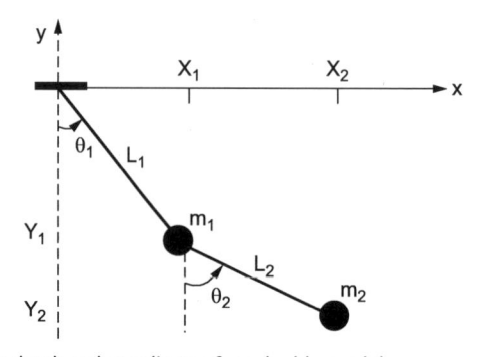

FIGURE 4.199 Geometry sketch and coordinates for a double pendulum.

The Euler-Lagrange equations are:

$$\frac{\partial L}{\partial \theta_1} - \frac{d}{dt}(\frac{\partial L}{\partial \dot{\theta}_1}) = 0$$

$$\frac{\partial L}{\partial \theta_2} - \frac{d}{dt}(\frac{\partial L}{\partial \dot{\theta}_2}) = 0$$

We now apply Euler-Lagrange equations using the resulting equation for L. The final result is a system of nonlinear ODEs, as given below:

$$(m_1 + m_2)L_1^2\ddot{\theta}_1 + m_2L_1L_2\ddot{\theta}_2\cos(\theta_1 - \theta_2) + m_2L_1L_2\dot{\theta}_2^2\sin(\theta_1 - \theta_2)$$
$$+(m_1 + m_2)L_1 g\sin\theta_1 = 0$$

$$m_2L_2^2\ddot{\theta}_2 + m_2L_1L_2\ddot{\theta}_1\cos(\theta_1 - \theta_2) - m_2L_1L_2\dot{\theta}_1^2\sin(\theta_1 - \theta_2) + m_2L_2 g\sin\theta_2 = 0$$

Solution

To solve the nonlinear system of ODEs, we let m_1 = *50 g*, m_2 = *100 g*, L_1 = *20 cm*, L_2 = *10 cm*. The initial conditions are $\theta_1 = \dot{\theta}_1 = \dot{\theta}_2 = 0$ and $\theta_2 = \pi/6$.

1. Launch COMSOL and open a new file. Save it as Example 4.23.

2. Select 0D from the Model Wizard window and click Next ➪.

3. From the Add Physics list, expand the Mathematics node and select ODE and DAE Interfaces>Global ODEs and DAEs (*ge*). Click Next ➪.

4. From the Select Study Type list, select Time Dependent and click Finish 🏁.

For this example, we don't have geometry to build (hence the 0D option selected). We define the equations using the Model 1 (*mod1*) node in the Model Builder window.

5. Expand the Model 1 (*mod1*) node in the Model Builder window and click on the Global Equations 1 node. In the corresponding window, enter the governing equations derived for θ_1, θ_2, and their initial conditions. See Figure 4.200. Note that the variable names should be entered under column Name, and they should be consistent with the variables used in the equations. We used **theta1** for θ_1 and **theta2** for

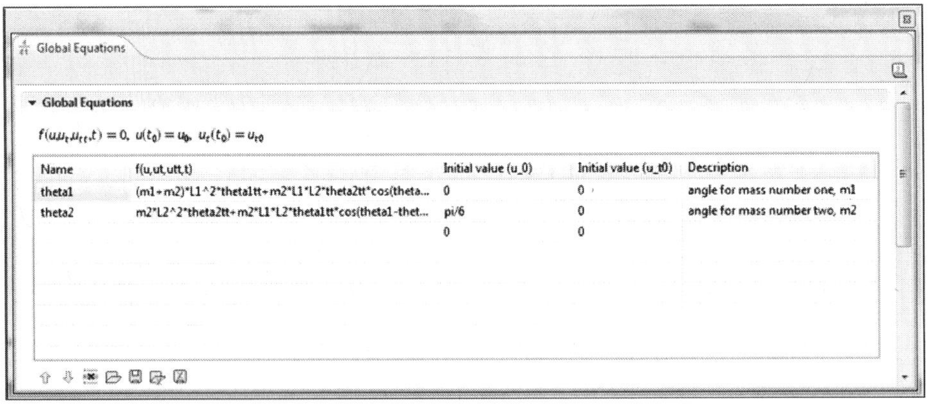

Global Equations

▾ Global Equations

$f(u, u_t, u_{tt}, t) = 0,\ u(t_0) = u_0,\ u_t(t_0) = u_{t0}$

Name	f(u,ut,utt,t)	Initial value (u_0)	Initial value (u_t0)	Description
theta1	(m1+m2)*L1^2*theta1tt+m2*L1*L2*theta2tt*cos(theta...	0	0	angle for mass number one, m1
theta2	m2*L2^2*theta2tt+m2*L1*L2*theta1tt*cos(theta1-thet...	pi/6	0	angle for mass number two, m2
		0	0	

FIGURE 4.200 Entering system of ODEs, as Global equation.

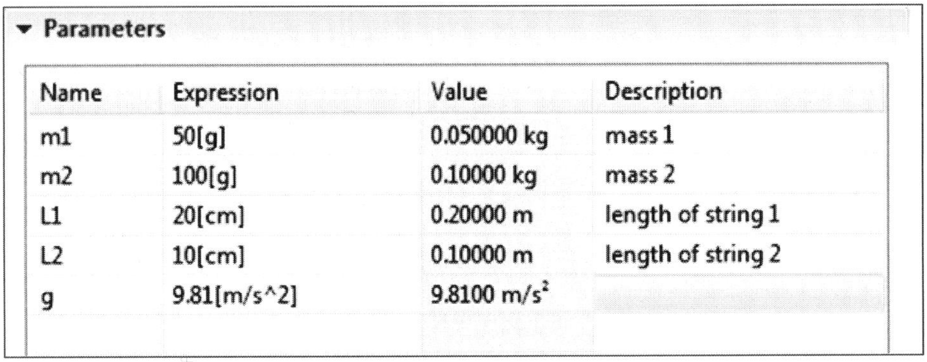

▾ Parameters

Name	Expression	Value	Description
m1	50[g]	0.050000 kg	mass 1
m2	100[g]	0.10000 kg	mass 2
L1	20[cm]	0.20000 m	length of string 1
L2	10[cm]	0.10000 m	length of string 2
g	9.81[m/s^2]	9.8100 m/s²	

FIGURE 4.201 Data for the double pendulum, as Parameters.

θ_2. Note that in COMSOL derivatives are available. For example, the time derivative of **theta1** is **theta1t,** and so on.

6. To enter the data, we enter them as parameters into the model. Right-click on the Global Definitions node in the Model Builder window and select Parameters. In the corresponding window, enter the data as shown in Figure 4.201.

7. Expand the Study 1 node in the Model Builder window and click on Step 1: Time Dependent. In the corresponding window, enter **range (0, 0.05, 4).**

8. Click on Study 1 and select Compute. Default results show line graphs for theta1 and theta2, as shown in Figure 4.202.

9. It is useful to calculate the variations of, for example, x-displacements of mass1 and mass2, as well. Click on the Global 1 node and in the corresponding window find the Data section. In the table, type **L1*sin(mod1.theta1)** in the first row and **L1*sin(mod1.theta1)+L2*sin(mod1.theta2)** in the second row. These are expressions for X_1 and X_2, respectively. Click on the Plot icon. Results are shown in Figure 4.203.

Similarly, variations of y-coordinates of the two masses could be calculated.

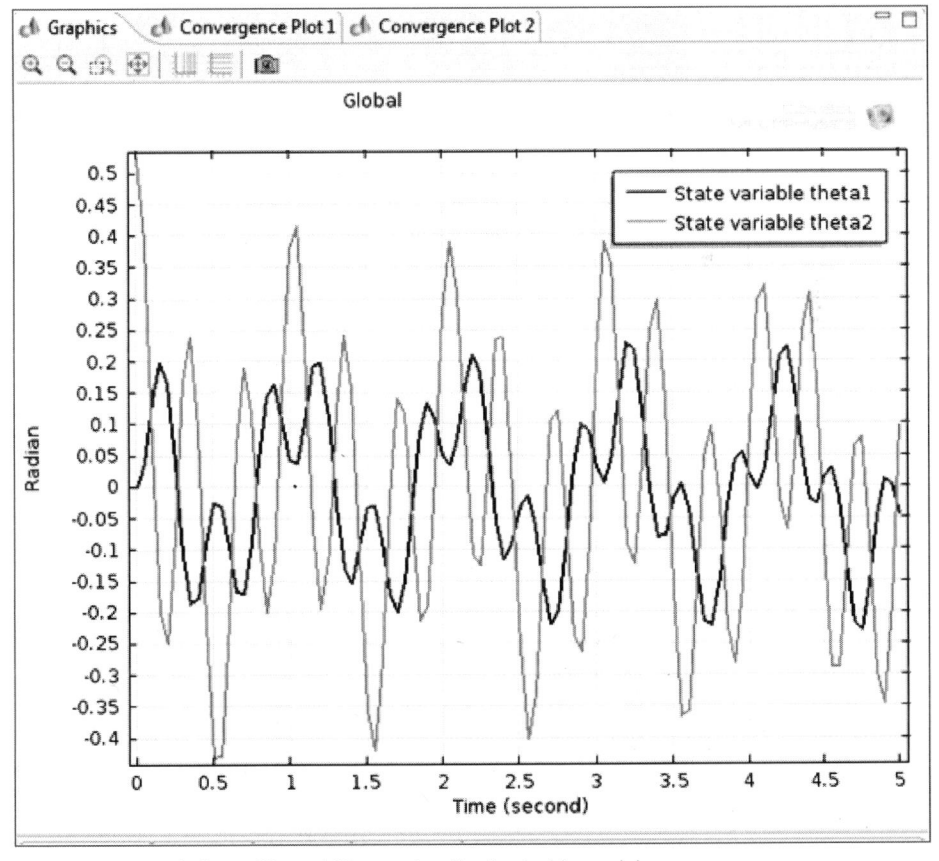

FIGURE 4.202 Variations of θ1 and θ2 w.r.t. time for the double pendulum.

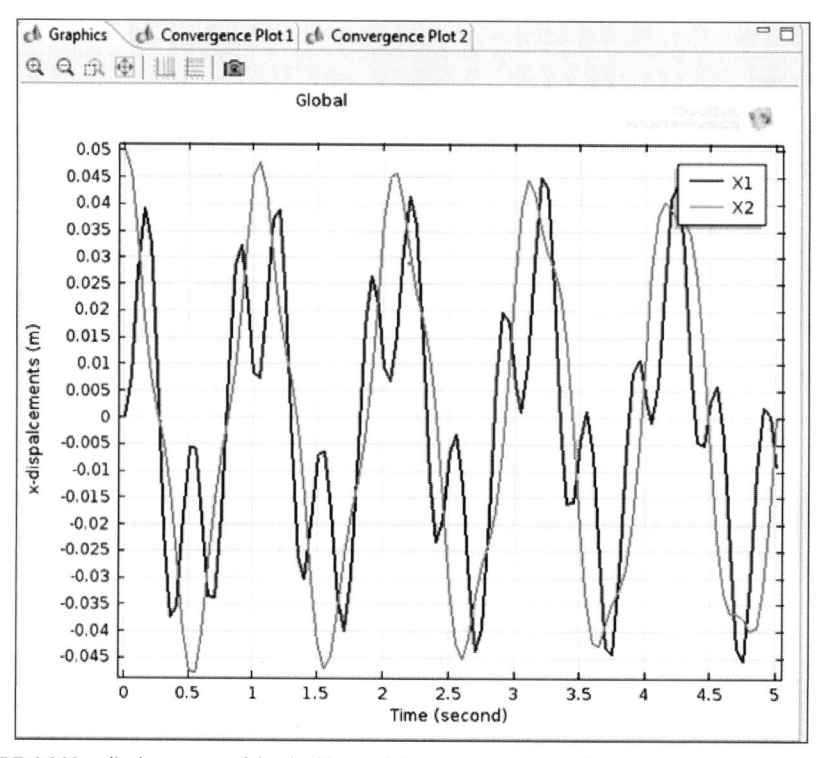

FIGURE 4.203 x-displacements of the double pendulum mass1 and mass2 w.r.t. time.

It is well-known that double pendulums may exhibit periodic, quasi-periodic, or chaotic behavior depending on the amount of initial energy input to the system. Double pendulum is a complex system. In order to analyze its dynamics, several graphs usually are needed (e.g., velocity vs. position, trajectories, angular velocities, angles). COMSOL allows users to make graphs of these quantities.

10. Click on the Global1 node (under 1D Plot Group). In the corresponding window, find and expand the Data section. Expand y-Axis Data and enter **mod1.theta1** in the first row under Expression. Expand x-Axis Data and from the Parameters list choose Expression and enter **mod1.theta2.** Expand the Coloring and Style section and select Point for Marker and In data Points for Positioning. Enter the axes and graph title, if desired, and click Plot. Results are shown in Figure 4.204. As shown in this graph, when mass2 is along the y-axis, hence theta2 = 0, mass1 is at a different known location on the x-y plane. From this graph users may extract the locations of mass1 and mass2 for different values of theta1 and theta2.

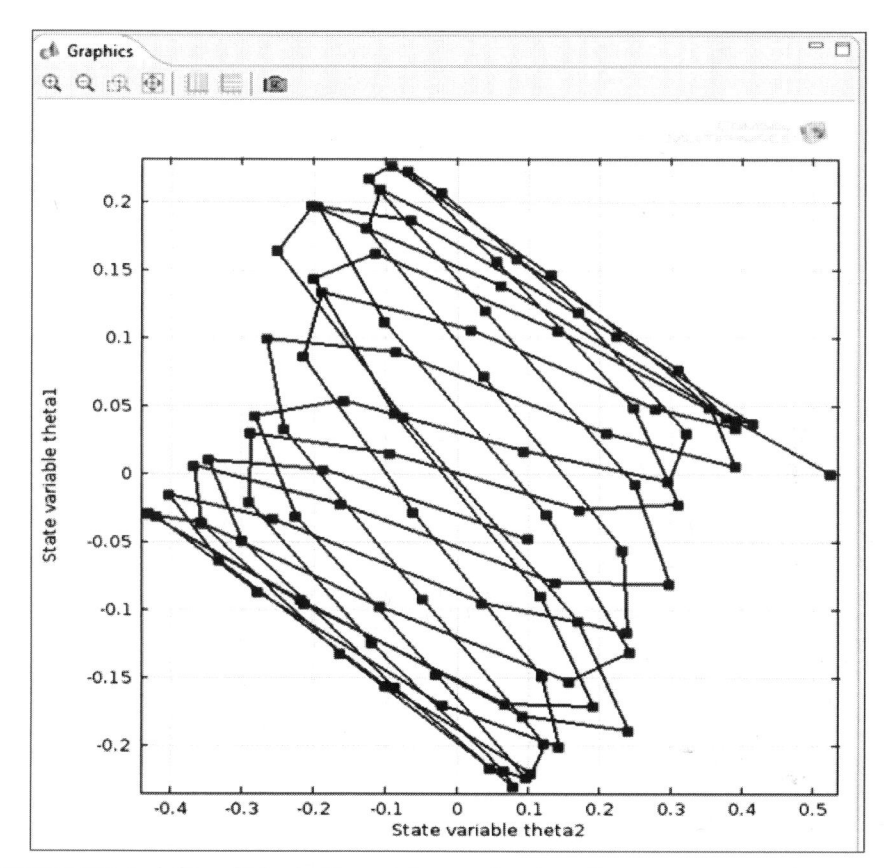

FIGURE 4.204 Angle θ1 vs. θ2 for double pendulum mass1 and mass2.

Example 4.24: Multiphysics Model for Thermoelectric Modules

Thermoelectric (TE) materials can be used for power generation and refrigeration, as shown in Figure 4.205. In this example, we model a thermoelectric module system. The TE modules in a system are connected in series and in parallel to produce useful power and required electrical voltage and current.

When two dissimilar materials are connected together (electrically in series and thermally in parallel) and a differential temperature is applied at their two ends, an electric voltage is generated. This is known as the Seebeck effect, which states that generated voltage is proportional to the temperature difference. When an electrical load is connected to the module, then electric current running through generates electric power. There are also other effects, such as Peltier and Thomson effects, associated with current going through thermoelectric module (see Reference 4.14).

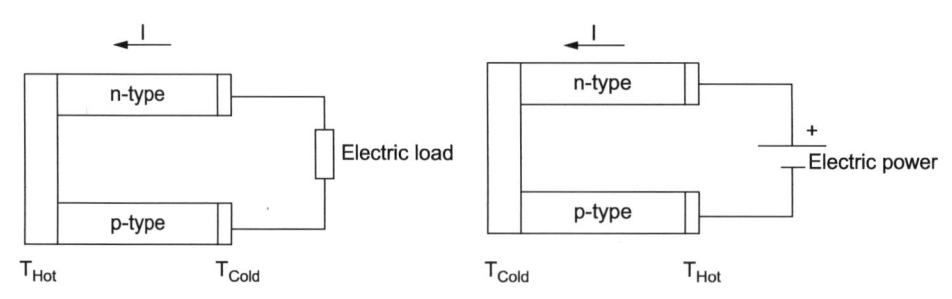

FIGURE 4.205 Seebeck effect: electric power generated by providing $\Delta T = T_{Hot} - T_{Cold}$ (left diagram) and Peltier effect: refrigeration by absorbing heat on the T Cold side providing electric power input (right diagram).

The Peltier effect appears when an external electric power is applied instead of the electric load and as a result a temperature difference occurs, as shown in Fig. 4.205. Thermoelectric refrigeration uses the Peltier effect.

The efficiency of a TE module (e.g., related to power generation mode) is referred to as figure-of-merit. The figure-of-merit is given as, $Z = \dfrac{S^2 \sigma}{k}$ where S is the Seebeck coefficient (V/K), σ electrical conductivity ($\Omega^{-1} m^{-1}$), and k thermal conductivity (W/m.K). The product $S^2 \sigma$ is referred to as the electrical power factor. The figure-of-merit for a TE device is sometimes given in its dimensionless form, as ZT, where T is absolute temperature.

The main phenomena associated with a TE are heat flux and electric current density. Heat flux is proportional to temperature gradient (Fourier's law) and generates electric current (Seebeck effect). Electric current by itself heats up the materials (Joule heating effect). The electric current absorbs heat reversibly (Peltier effect). A combination of these phenomena constitutes the governing equation for heat transfer in a TE (in vector/tensor format):

$$q = \Pi \cdot J - k \cdot \nabla T$$
$$\nabla \cdot q = \dot{Q}$$

where q is heat flux (W/m^2), Π Peltier coefficient matrix (V), J current density (A/m^2), and T temperature (K). $\dot{Q} = J \cdot E$ is the heat source due to the Joule heating effect and $E = -\nabla V$ is the electric field due to electric voltage potential V.

The electric current density is due to Seebeck voltage and should satisfy Ohm's law. Hence we have the governing equation for current density as:

$$J = \sigma \cdot (E - S \cdot \nabla T)$$
$$\nabla \cdot J = 0$$

In the last equation ($\nabla \cdot J = 0$), which governs convergence of current density, we neglect electric displacement (or time variation of electric flux density). This assumption is valid for TE.

By combining and manipulating the above equations (see References 4.15 and 4.16), we have:

$$\begin{cases} \nabla \cdot [-(S^2 \sigma T + k) \nabla T] + \nabla \cdot (-S \sigma T \nabla V) = \sigma[(\nabla V)^2 + S \nabla T \cdot \nabla V)] \\ \nabla \cdot (S \sigma \nabla T) + \nabla \cdot (\sigma \nabla V) = 0 \end{cases}$$

These equations are the mathematical model for TE. Note that these are coupled nonlinear partial differential equations in vector notation format.

We use the Mathematics module in COMSOL for modeling, specifically the Coefficient Form for solving the above equations.

In COMSOL definition, the Coefficient Form of a PDE reads:

$$e_a \frac{\partial^2 u}{\partial t^2} + d_a \frac{\partial u}{\partial t} + \nabla \cdot (-c \nabla u - \alpha u + \gamma) + \beta \cdot \nabla u + au = f$$

where u is the dependent variable vector and coefficients are defined independently. Therefore, after comparing our mathematical model with the Coefficient Form, we have:

$$u = \begin{pmatrix} T \\ V \end{pmatrix} \quad , \quad c = \begin{pmatrix} k + S^2 \sigma T & S \sigma T \\ S \sigma & \sigma \end{pmatrix} \quad , \quad f = \begin{pmatrix} \sigma[(\nabla V)^2 + S \nabla T \cdot \nabla V)] \\ 0 \end{pmatrix}$$

All other coefficients (e_a, d_a, α, β, γ, a) are zero. We will discuss the boundary conditions during model construction.

Solution[4]

1. Launch COMSOL and open a new file. Save it as Example 4.24.

2. Select 3D from the Model Wizard window and click Next ➪.

---

[4] Version 4.4 has a thermoelectric modeling module built in.

3. From the Add Physics list, expand the Δu Mathematics node and select Δu PDE Interfaces>Δu Coefficient Form PDE(*c*). Click on the Δu Coefficient Form PDE (*c*) and then click Add Selected ✚. Locate the Dependent variables section and enter **2** in the space for Number of dependent variables and **T** and **V** for Dependent variables. Leave Field name as default. Click Next ➪.

4. From the Select Study Type list, select Stationary and click Finish 🏁.

5. To build the TE modules geometry, right-click on the Geometry 1 node in the Model Builder window and select Block. In the corresponding window, enter 1.24[cm] for Width and 1[cm] for both Depth and Height. For Position enter 0.1[cm] for z. Click the Build Selected icon on the toolbar of the Block window. Similarly, build more blocks with the dimensions provided in Table 4.6.

6. Right-click on the Geometry 1 node and select Transforms>Copy. In the corresponding window, add Blocks 1, 3, and 4 to the selection list. In the Displacement section, enter 3.24[cm] for x: and click on Build Selected. The final result for the geometry is shown in Figure 4.206.

7. We use the Parameters feature for entering material properties. As mentioned, p-type (Block 1) and n-type (Block 4) silicon-based materials are common in industry for TE modules. In addition, electrodes (Blocks 2 and 3) are usually made out of copper. Right-click on the Global Definitions node in the Model Builder window and select Parameters. In the corresponding window, enter the data as shown in Figure 4.207. These are typical material properties for Bismuth-Telluride and copper.

For this example, we consider constant material properties. However, the model is capable of using materials with varying properties, such as functions of temperature.

	Width	**Depth**	**Height**	**Position (x, y, z)**
Block 2	1.24[cm]	1[cm]	0.1[cm]	(0, 0, 0)
Block 3	2.74[cm]	1[cm]	0.1[cm]	(0, 0, 1.1[cm])
Block 4	1[cm]	1[cm]	1[cm]	(1.74[cm], 0, 0.1[cm])
Block 5	2.74[cm]	1[cm]	0.1[cm]	(1.74[cm], 0, 0)
Block 6	1[cm]	1[cm]	0.1[cm]	(4.98[cm], 0, 0)

TABLE 4.6 Dimensions for building geometry of TE modules.

COMSOL MODELS FOR PHYSICAL SYSTEMS • **223**

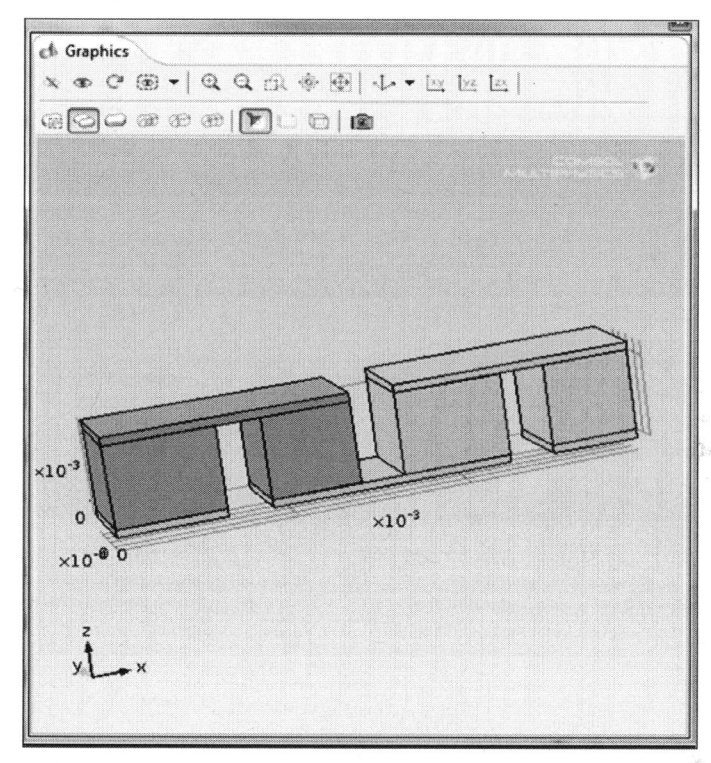

FIGURE 4.206 Built geometry for two TE modules connected in series.

FIGURE 4.207 Parameters for material properties for TE modules, p-type, n-type, and copper electrodes.

8. To enter the model equations, click on the Coefficient Form PDE 1 node in the Model Builder window. Rename this node to **Coefficient Form for TEp**. In the corresponding window, expand the Equation section and locate the PDE and boundary conditions, in terms of COMSOL coefficient form. We will enter expressions for Coefficients c and f only and leave the rest as default, as discussed previously. Expand the Diffusion Coefficient section and enter the expressions for c and f matrices for the p-type TE (Domains 2 and 6), as shown in Figure 4.208. The variables should be exactly the same as those defined as Parameters (see Figure 4.207) and they are case sensitive. Users should note that variables are made dimensionless by applying the inverse of their corresponding units. The color of expressions entered in the space holders should turn black when units are consistent. The dimensions of the elements of the coefficient c entered are m^2 in order to have the entire equation dimensionless after gradient operators are applied twice, i.e., the term $\nabla \cdot (-c\nabla \mathbf{u})$.

9. Similarly, we define coefficients for n-type TE modules. Right-click on the Coefficient Form PDE (c) node in the Model Builder window, and select Coefficient Form PDE from the list. Coefficient Form PDE 2 node will appear; rename it to **Coefficient Form for TEn**. In the corresponding window, select and add domains 5 and 9 to the list under Domain Selection. Expand the Diffusion Coefficient section and enter

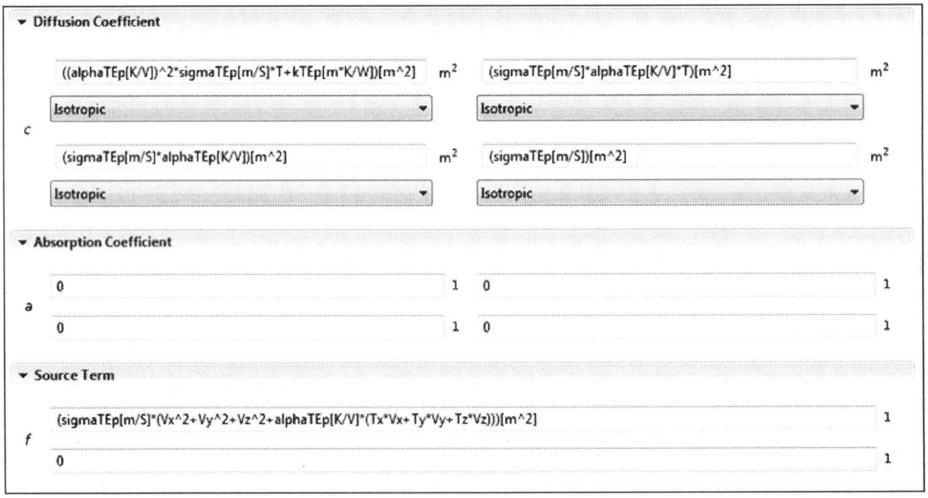

FIGURE 4.208 Coefficient expressions for p-type TE blocks.

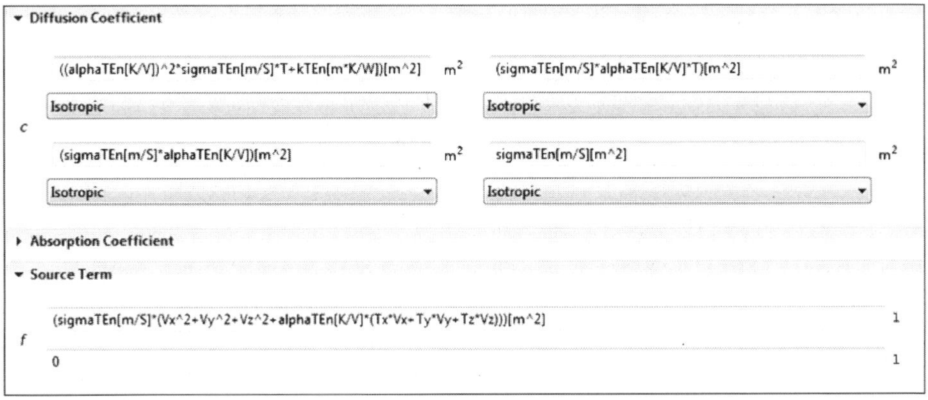

▼ Diffusion Coefficient			
c	((alphaTEn[K/V])^2*sigmaTEn[m/S]*T+kTEn[m*K/W])[m^2] m²	(sigmaTEn[m/S]*alphaTEn[K/V]*T)[m^2] m²	
	Isotropic ▼	Isotropic ▼	
	(sigmaTEn[m/S]*alphaTEn[K/V])[m^2] m²	sigmaTEn[m/S][m^2] m²	
	Isotropic ▼	Isotropic ▼	
▶ Absorption Coefficient			
▼ Source Term			
f	(sigmaTEn[m/S]*(Vx^2+Vy^2+Vz^2+alphaTEn[K/V]*(Tx*Vx+Ty*Vy+Tz*Vz)))[m^2] 1		
	0 1		

FIGURE 4.209 Coefficient expressions for n-type TE blocks.

the expressions for c and f matrices for the n-type TE (Domains 5 and 9), as shown in Figure 4.209. The variables should be exactly the same as those defined as Parameters (see Figure 4.207) and they are case sensitive. Nondimensionalization of the expressions for coefficient c is applied similar to Step 9 above.

10. Similarly, we define coefficients for copper electrodes. Right-click on the Coefficient Form PDE (c) node in the Model Builder window, and select Coefficient Form PDE from the list. Coefficient Form PDE 3 node will appear; rename it to **Coefficient Form for Copper Electrodes**. In the corresponding window, select and add domains 1, 3, 4, 7, and 8 to the list under Domain Selection. Expand the Diffusion Coefficient section and enter the expressions for c and f matrices for the electrodes, as shown in Figure 4.210. The variables should be exactly the same as those defined as Parameters (see Figure 4.207) and they are case sensitive.

11. This model can be used for modeling Seebeck-type thermoelectric power generation or Peltier-type thermoelectric refrigeration. For this example we apply a set of boundary conditions for power generation. Therefore we should apply temperatures at the electrodes to calculate the resulting electrical voltage. Right-click on the Coefficient Form PDE (c) node and select Constraint. Constraint 1 node will appear in the model tree list. Select boundaries corresponding to electrodes on the top of the modules (i.e., 10, 35) from the geometry and add them to the Selection list in the corresponding window. Expand the Equation section. From the equation we can see that we should define

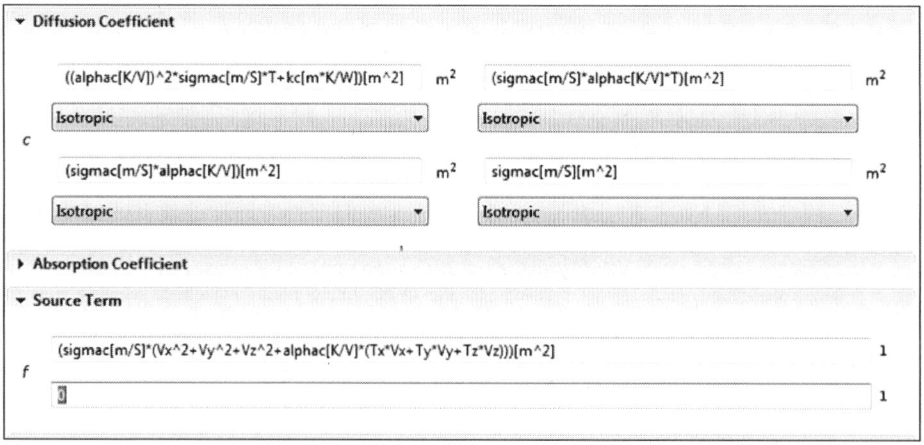

FIGURE 4.210 Coefficient expressions for copper electrode blocks.

values for R, corresponding to the $0 = R$ equation. In the Constraint section, enter **180-T** and **0** into values for R. This will set the temperature at the upper electrodes at 180° C. Similarly, create another constraint boundary condition for the remaining electrode surfaces (i.e., boundaries 3, 19, 43) and enter **10–T** and **–V** into the values for R. This will set the temperature at the lower electrodes at 10° C and ground the voltage for reference at zero.

12. Now the model is ready for meshing. Click on Mesh 1 and in the corresponding window, accept default inputs and click on Build All.

13. To run the model, click on the Study 1 node, and then click on Compute in the toolbar.

14. The results appear in the Graphics window. The default is the Slice 1. Click on Slice 1 and in the corresponding window under the Expression, type T and check the results. Then type V and check the results. Number of slices and their orientations can be modified by changing the parameters in the Plane Data section. Other variables, such as electric field, could be visualized by typing $V_x + V_y + V_z$ for the Expression. To add another graph, right-click on the Results node in the Model Builder and select 3D Plot Group. The 3D Plot Group 2 node will appear in the model tree; right-click on it and select Surface. Surface 1 node will appear; in the corresponding window enter the desired variable in the Expression space, such as V for visualization. See Figure 4.211.

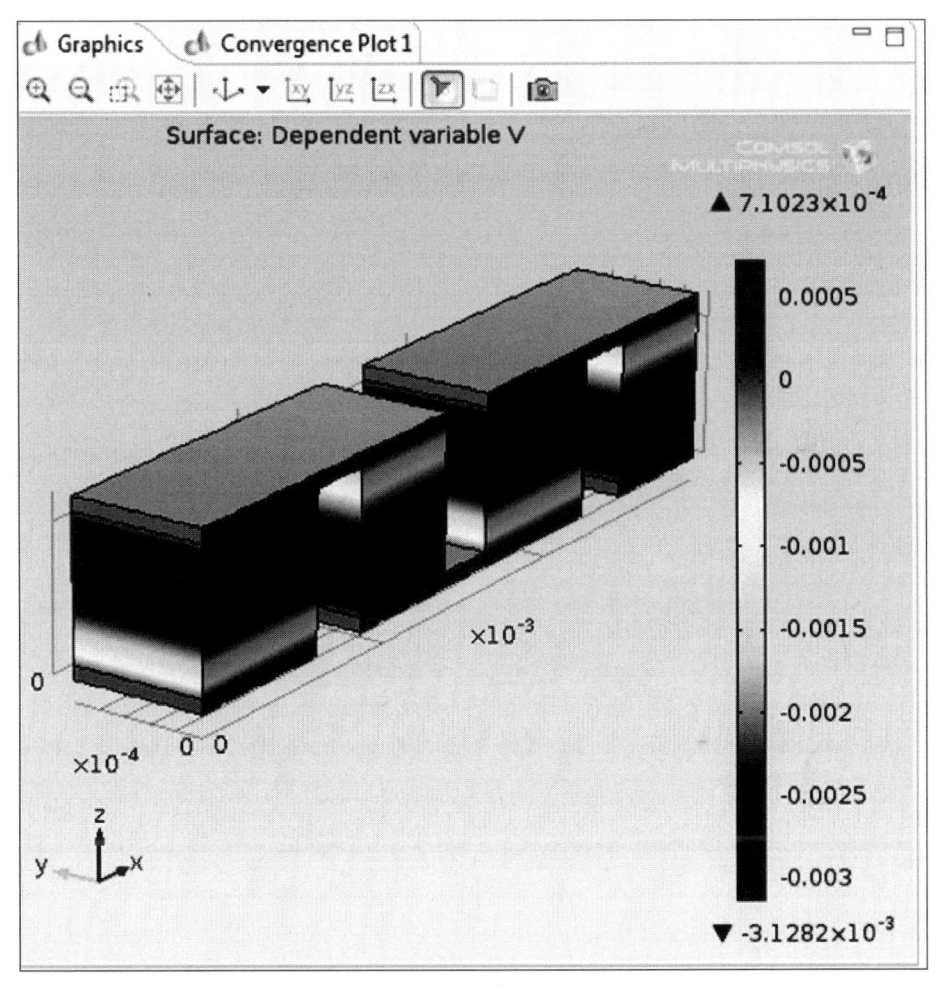

FIGURE 4.211 Results for voltage variations for TE modules.

The numerical values for voltage, after validation, could be used for generated power. Validation for this model could be performed when experimental results are available. We leave this task open for those readers who may have access to the experimental setup. It can be shown that changing the temperature at the upper electrode surfaces will change the resulting voltage generated. This will suggest that we could use temperature-dependent material properties to get more accurate results. Readers are referred to Reference 4.17 for the data and further reading.

Example 4.25: Acoustic Pressure Wave Propagation in an Automotive Muffler

In this example, we use the COMSOL Multiphysics module to model and analyze pressure wave propagation and attenuation in a muffler.

Sound travels through air as a longitudinal wave, that is, the material (for example, air molecules) displacement and the sound wave propagation are in the same direction. When a sound wave is created by a source, such as a speaker or human vocal cords, the extra pressure due to compression of air molecules propagates as a wave. The extra pressure or acoustic pressure is in addition to the background ambient pressure or atmospheric pressure in the air or the medium involved. It is important to understand that the velocity of the air molecules is different from the velocity of the sound wave generated. Speed of sound wave V_s in a medium is, in general, proportional to the square root of medium stiffness over its density. For air as an ideal gas, $V_s \cong 20\sqrt{T}$ where T is the absolute temperature of air. Typical sound wave velocity at room temperature is 343 m/s depending on the source; sound wave length λ and frequency f are defined and their product should be equal to the generated sound wave velocity $V_s = \lambda f$. Acoustic pressure, or the generated sound wave amplitude p_s depends on the energy released from the sound source. The flux of this energy (i.e., energy per unit area per unit time) is called sound Intensity I, which is also equal to acoustic power per unit area and is proportional to p_s^2. If r is the distance from the sound sources, then sound intensity variation is proportional to $1/r^2$ and p_s to $1/r$ (see Reference 4.18). The human ear, as a mechanical sound receiver device, is sensitive to sound frequency (20Hz to 20kHz) and can detect sound intensities as low as $I_0 = 10^{-12}$ W/m², which corresponds to a sound pressure equal to about 2×10^{-5} Pa. I_0 is used as a scale to define sound level or "loudness" equal to $10\log_{10}(I/I_0)$ as decibels and acoustic sound pressure level (SPL). Humans' hearing comfort sound level is about 50 dB.

To model the sound pressure-wave propagation as a result of combustion in a car engine, through a muffler we use the COMSOL model library and relevant data, after some modifications. For more details and mathematical models refer to Acoustics_Module/Industrial_Models/absorptive_muffler[5]. The main objective of a muffler is to attenuate the sound level to a bearable level for human hearing. In this model, the muffler is considered as a resonator with empty space.

---

[5] Model made using COMSOL Multiphysics® and is provided courtesy of COMSOL. COMSOL materials are provided "as is" without any representations or warranties of any kind including, but not limited to, any implied warranties of merchantability, fitness for a particular purpose, or noninfringement.

Solution

1. Launch COMSOL and open a new file. Save it as Example 4.25.
2. Select 3D from the Model Wizard window and click Next ⇨.
3. From the Add Physics list, expand Acoustics and select Pressure Acoustics>Pressure Acoustics, Frequency Domain (*acpr*), and then click Add Selected ✚. Locate Dependent variables and notice that Pressure p is defined as the sound pressure. Click Next ⇨.
4. From the Select Study Type list, select Frequency Domain and click Finish 🏁.
5. To define the model data as parameters, right-click on the Global Definitions and select Parameters. In the corresponding window, enter the data shown in Figure 4.212. Alternatively, users can use Load from File tool and import the file **absorptive_muffler_parameters.txt,** usually located in the Models folder where COMSOL is installed.

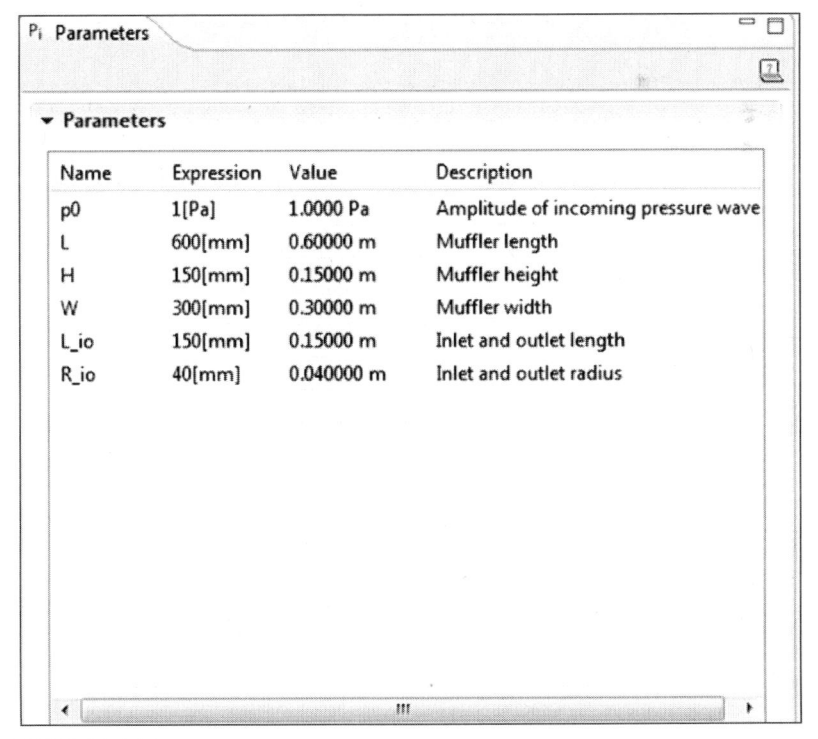

Name	Expression	Value	Description
p0	1[Pa]	1.0000 Pa	Amplitude of incoming pressure wave
L	600[mm]	0.60000 m	Muffler length
H	150[mm]	0.15000 m	Muffler height
W	300[mm]	0.30000 m	Muffler width
L_io	150[mm]	0.15000 m	Inlet and outlet length
R_io	40[mm]	0.040000 m	Inlet and outlet radius

FIGURE 4.212 Model data as parameters.

To build the geometry, we use the CAD tools available in COMSOL. Users may want to build it using their desired CAD software.

6. Click on the Geometry 1 node and in the corresponding window change Length to **mm**. Right-click on the Geometry 1 node and select Cylinder from the list. In the corresponding window, locate the Size and Shape section and enter **R_io** for Radius and **L_io** for Height. For Position, enter **–L_io**. In the Axis section, select Cartesian from the list for Axis type. Enter 1 for x. Then click Build Selected.

7. Similarly, build another cylinder. All data is the same except the Position x value should be **L**.

These two cylinders are the inlet and outlet pipes to the muffler resonate chamber. We now build this chamber.

8. Right-Click on Geometry 1 and select Work Plane. In the corresponding window, select zy-plane from the Plane section. Click Build Selected. Right-click on the Plane Geometry node (created under Work Plane 1) and select Ellipse. In the corresponding window, enter **H/2** for a-semiaxis, **W/2** for b-semiaxis, and 360 for Sector angle. Click Build Selected. Right-click on Work Plane 1 and select Extrude. In the corresponding window, enter **–L** for Distances, under Distances from Plane section. Click Build All and then click on the Form Union (*fin*) node, in the Model Builder window. Final geometry is shown in Figure 4.213.

9. We now add material properties of the air inside the muffler. Right-click on the Materials node and select Open Material Browser. In the corresponding window, select Liquid and Gases>Gases>Air and add it to the model by right-clicking. By default, air is added to all three domains (1, 2, 3) for the muffler geometry.

10. Click on the Pressure Acoustic Model 1 node. In the corresponding window, extend Equation and study the equation. This equation can be derived from a general wave equation using a transform, like Fourier, to change the time domain to frequency domain. The result is a "Helmholtz" type equation that governs the acoustic pressure-wave propagation for different frequencies as shown in the Equation section. Variable p is a function of space and angular velocity ω (which is itself a function of frequency). We accept all default values given under the Model Inputs and Pressure Acoustics Model sections.

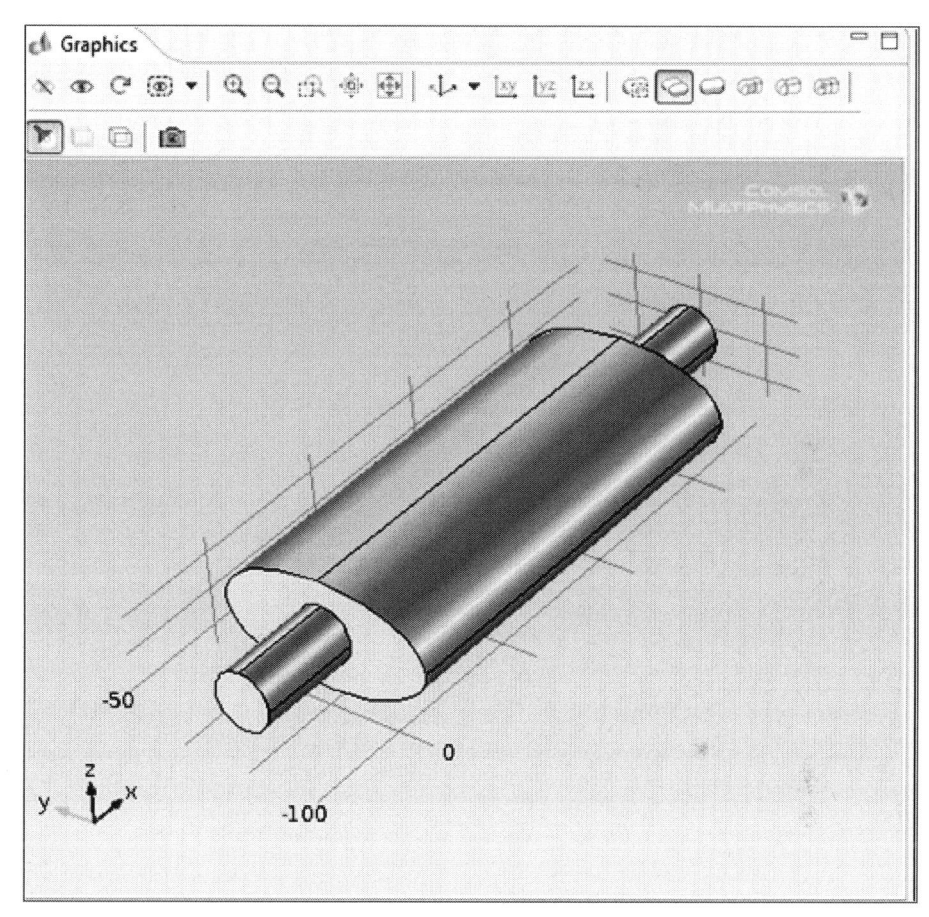

FIGURE 4.213 Graphics window showing muffler built geometry.

11. For the boundary conditions at the inlet and outlet, we have incoming and outgoing plane sound waves. To assign boundary conditions, we first assign explicit names for them. Right-click on the Definitions node located under Model 1 (*mod1*) in the Model tree, and select Selections>Explicit. In the corresponding window, expand Boundary from the list for Geometric entity level and select and add boundary 1. Right-click on the Explicit 1 node and rename it Inlet. Similarly, create Explicit 2 and rename it to Outlet and assign boundary 18 to it.

12. Right-click on the Pressure Acoustics, Frequency Domain (*acpr*) node and select Plane Wave Radiation. In the corresponding window, add

boundaries 1 and 18 (inlet and outlet surfaces of the pipes) by selecting them from the geometry in the Graphics window. Then right-click on the Plane Wave Radiation 1 node and select Incident Pressure Field. In the corresponding window, select Inlet from the list and add boundary 1 to the list. All other walls of the muffler have Hard Boundary. By assigning this type of boundary condition we assume that the normal derivative of the sound pressure is zero at the walls. Figure 4.214 shows the result for model tree nodes under Model 1(*mod1*), so far. The Variables 1a, Integration 1(*intop1*), and Integration 2(*intop2*) nodes will be built later in the process.

13. Although automatic tetrahedral mesh would work, we would like to build a custom mesh with maximum element size of one-fifth of the minimum sound wave length, which corresponds to the maximum frequency. This value is equal to 343/(1500*5). The number 343 is the sound velocity in *m/s* and 1500 Hz is the maximum frequency considered for this problem. Expand the Mesh 1 node and click on Size,

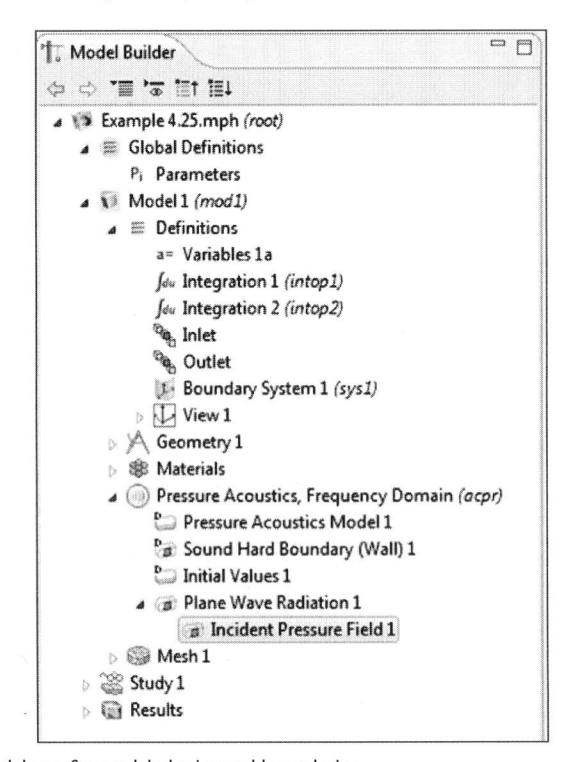

FIGURE 4.214 Model tree for model physics and boundaries.

and in the corresponding window click the Custom button. Enter **343[m/s]/1500[Hz]/5** for the Maximum element size. Right-click on Mesh 1 and select More Operations> Free Triangular. In the corresponding window, choose Boundary from the list for Geometric entity level. Select boundaries 6 and 9 and add them to the list. These are boundaries at the face of the muffler resonator at the end of the entrance/inlet pipe. Click Build Selected. You may want to click on the Transparency node in the Graphics window toolbar to clearly see these surfaces, i.e., boundaries 6 and 9.

14. We build the volume elements using the Swept tool. This method is suitable for building a mesh for wave propagation. Right-click on Mesh 1 and select Swept. In the corresponding window, select and add Domain 1 (the inlet pipe) to the list and click Build Selected. Similarly, create Swept 2. In the corresponding window, select and add Domain 2 to the Selection list. In the Source Faces section, select and add boundaries 6 and 9. In the Destination Faces section, select and add boundaries 12 and 13 to the list. Click Build Selected. Finally, create Swept 3. In the corresponding window, click Build All. Figure 4.215 shows the final mesh.

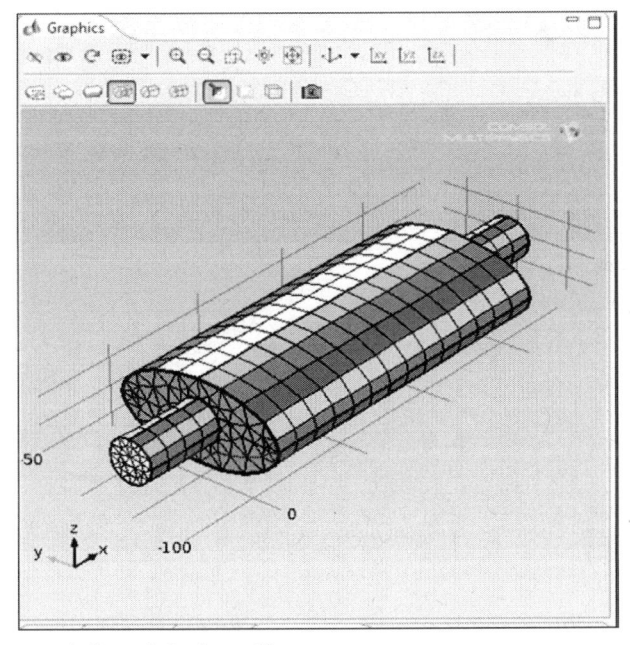

FIGURE 4.215 Custom built mesh for the muffler geometry.

15. To set the range for frequency, click on Step 1: Frequency Domain node under Study 1 in the Model Builder window. In the corresponding window, enter **range(50,25,1500)** in the Frequencies **value**. Right-click on the Study 1 node and select Compute. Wait until calculations are finished.

16. The default results for acoustic pressure appear in the Graphics window. Expand the Acoustic Pressure (*acpr*) node and click on Surface 1. In the corresponding window, locate the Expression section. Change the variable to **acpr.absp** by clicking on Replace Expression and select Pressure Acoustics, Frequency Domain>Absolute pressure>acpr.absp. Click Plot. The results, which are absolute (norm) values of the sound pressure on the surface of the muffler, appear in Graphics window, as shown in Figure 4.216. Graphs for other frequencies could be obtained by clicking on the Acoustic Pressure (*acpr*) node and choosing the desired value from the Parameter value (*freq*).

17. Click on the Sound Pressure Level (*acpr*) node. The graph appears in the Graphics window. The sound pressure is the sound pressure associated with acoustic pressure with reference to the 0 dB, associated with the lowest human hearing sensitivity or 10^{-12} W/m², which is set as 20 µPa in COMSOL. The result is shown in Figure 4.217.

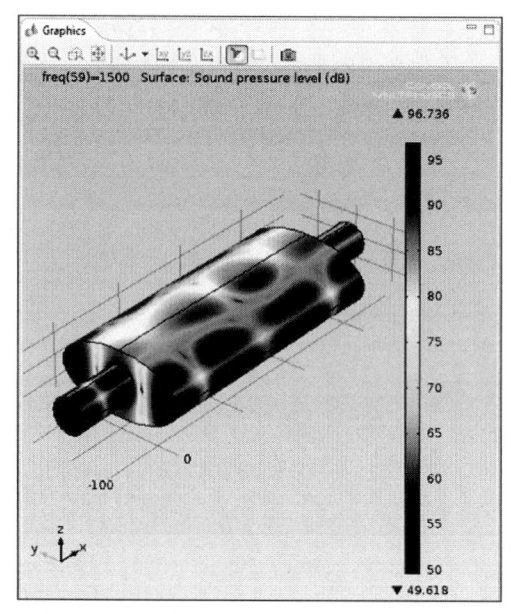

FIGURE 4.216 Total acoustic pressure on the surface of the muffler for frequency of 1500 Hz.

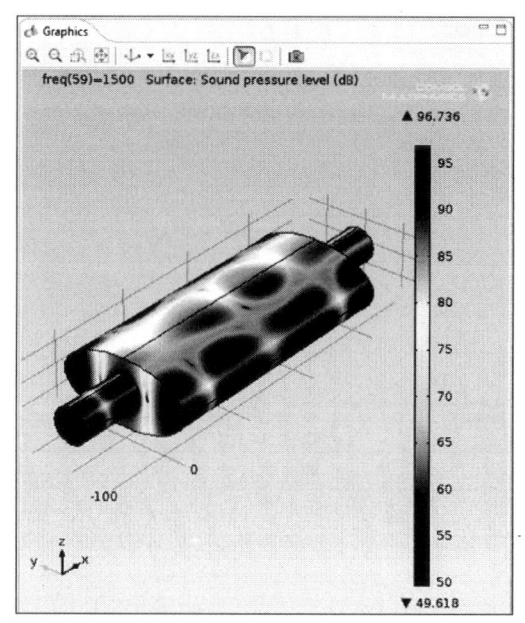

FIGURE 4.217 Sound pressure level (referenced to 0 dB) on the surface of the muffler for frequency of 1500 Hz.

18. Another useful graph is the isosurfaces for acoustic pressure, or locations for constant pressure in the muffler. Click on the Acoustic Pressure, Isosurfaces (*acpr*). We keep the default Expression acpr.p_t, which is the total acoustic pressure (real part). Click on Plot. Results are shown in Figure 4.218.

19. Attenuation of acoustic power is the main function of the muffler. Sound intensity or power per unit area is the quantity of interest given by $p^2/2\rho V_s$. To integrate this quantity over the surface area of the inlet and outlet pipes, we create two integral operations and define the intensity as a variable. Right-click on the Definitions node under Model 1(*mod1*), in the Model Builder window, and Select Variables. In the corresponding window, under Variables enter **W_in** for Name and intop1(p0^2/(2*acpr.rho*acpr.c)) for Expression. Similarly, in the second row, enter **W_out** for Name and **intop2(abs(p)^2/(2*acpr.rho*acpr.c))** for Expression. Then create the integral operations (*intop1* and *intop2*) by right-clicking on the Definitions node and select Model Couplings>Integration. In the corresponding window, select and add boundary 1 (if needed, change the Geometric entity level to Boundary). Similarly, create Integration 2 (*intop2*) and assign boundary 18 to it. Right-click on Study and select Compute to update all solutions.

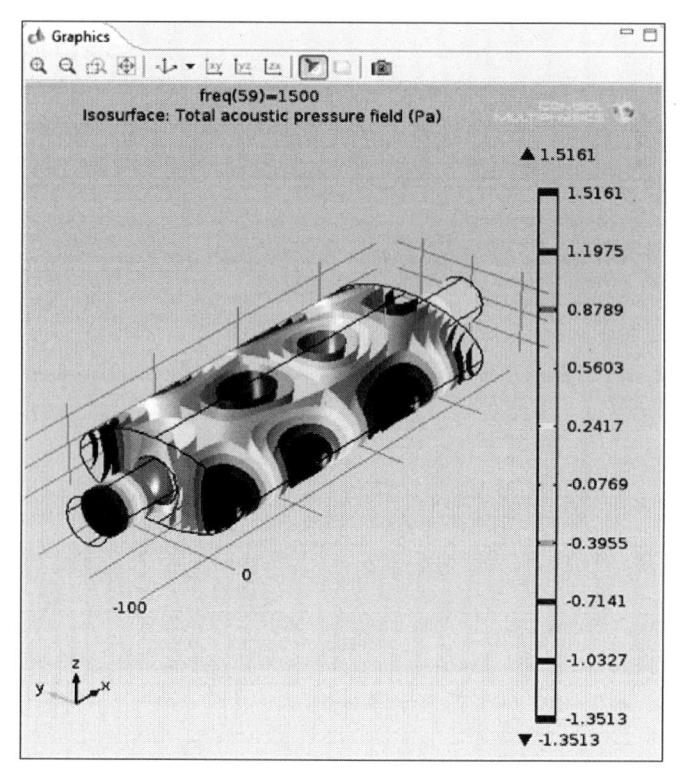

FIGURE 4.218 Total acoustic pressure isosurfaces on the surface of the muffler for frequency of 1500 Hz.

20. Now we are ready to extract the integrals of the acoustic power densities from the solution results. Right-click on the Results node and select 1D Plot Group. 1D Plot Group 4 appears in the model tree; right-click on it and select Global. In the corresponding window, locate the y_Axis data section and enter **10*log10(W_in/W_out)**. This expression is standard sound level in dB. It shows the ratio (logarithmic) of the incoming sound power over the outgoing one. Results are shown in Figure 4.219. Open Coloring and Style to change the graphing options for the line graph (2 for Width, Point for Marker, and In data points for Positioning).

The model is complete. Further studies or graphs may be built. In practice, a muffler resonate chamber is composed of two shells, instead of one, for damping the acoustic energy. Users may want to modify this model by adding another chamber inside the existing one and study its effects as compared to results obtained in this example.

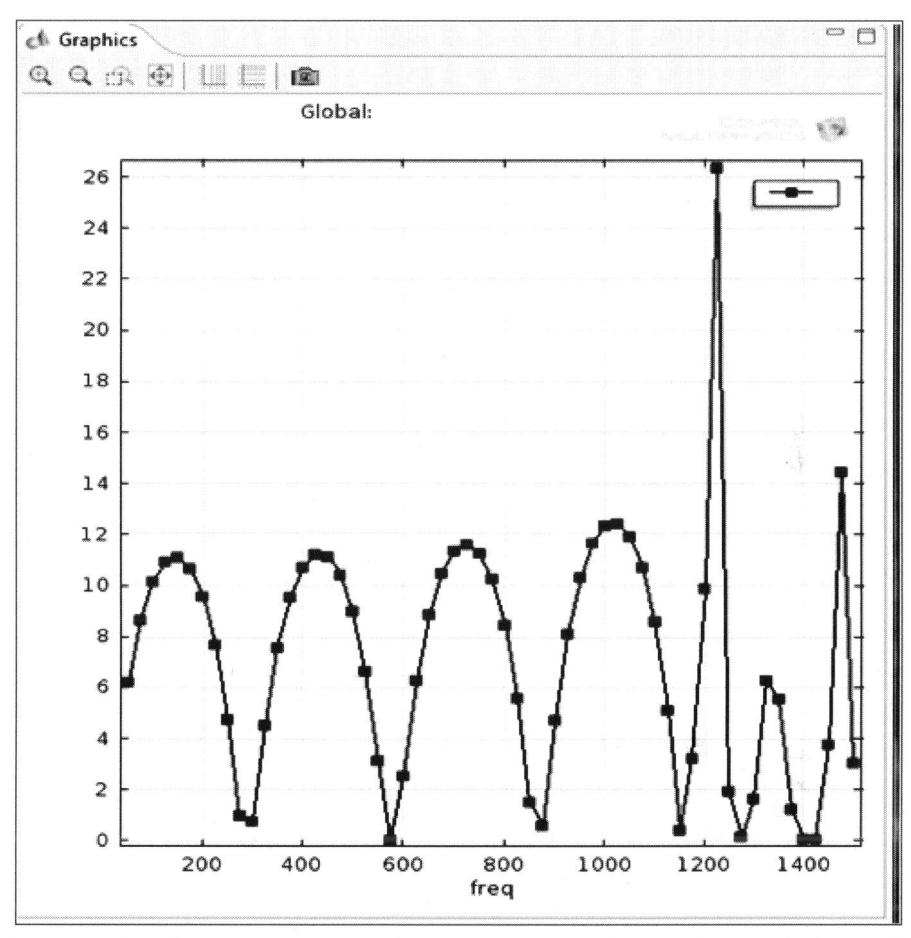

FIGURE 4.219 Decibel type acoustic power (in W) attenuation (10*log10(W_in/W_out)) of the muffler vs. frequency.

EXERCISE PROBLEMS

Problem 4.1: Using the solution to Example 4.1., show the principle stress directions by manipulating the results.

Problem 4.2: Repeat Example 4.1, using the symmetry of the geometry and loads along the line axis at y = 0.

Problem 4.3: Repeat Example 4.1 for different boundary conditions, for example, simply supported for vertical sides on the right side of the plate and triangular tension stress applied on the upper edge.

Problem 4.4: Calculate 10 natural frequencies of the plate given in Example 4.2.

Problem 4.5: Calculate the plate frequency given in Example 4.2, close to 1500 Hz, and show the corresponding modal shape.

Problem 4.6: Investigate the effects of different boundary conditions on the natural frequencies for the plate given in Example 4.2.

Problem 4.7: Repeat Example 4.3 for different sets of applied loads

Problem 4.8: Repeat Example 4.3 for different ranges of angle theta0.

Problem 4.9: Check the bracket stress given in Example 4.3 against Tresca yield criterion.

Problem 4.10: Calculate the moment of inertia of the column cross-section given in Example 4.4 and compare those obtained with the Euler's column formula $P_{cr} = \pi^2 EI / (4L^2)$.

Problem 4.11: Change the applied load given in Example 4.4 to higher values and make a graph of obtained critical load factors vs. applied loads.

Problem 4.12: For Example 4.4, assign different boundary conditions to the column and calculate buckling load. Compare the buckling load with and without lateral loads.

Problem 4.13: Modify the column given in Example 4.4 to have a variable cross-section (tapered). Calculate its critical load factor for 200 Pa compression load (use Scale 0.5 for x_w and y_w in the Extrude). Also in Example 4.4, change the values of point loads and repeat the Eigenfrequency calculations, both with and without loading.

Problem 4.14: In example 4.4, design the truss to have the optimum section to satisfy 0.6Fy (Fy is the yield strength equal to 240 MPa)

maximum applied stresses. Investigate the buckling for the member with maximum compression load.

Problem 4.15: In Example 4.4, introduce different damping to the truss vibrations and compare/discuss your results.

Problem 4.16: In Example 4.6, analyze the truss and compare your results with those in the model.

Problem 4.17: Repeat Example 4.6 for prescribed displacements, as boundary conditions, equal to 1 cm in x-direction and 2 cm in y-direction at point G.

Problem 4.18: Calculate the Reynolds number for the flow in Example 4.7.

Problem 4.19: For Example 4.7, run the model with finer meshes and discuss the results in terms of mesh dependency. Choose a physical point in the flow domain and investigate its numerical values, velocity, and pressure for different mesh sizes.

Problem 4.20: For a water-jet cutter, pressure of 20,000 psi is typical. Calculate the exit velocity for the nozzle using the model given in Example 4.7. Discuss validity of the assumptions for this model, such as water compressibility, turbulence, and other factors.

Problem 4.21: For a fly-wheel, the friction between the rotating disk and air is significant. Calculate the total forces applied on the disk due to shear stresses. Consider the dynamic viscosity of the air as a parameter and draw the total friction force versus viscosity. Use the model from Example 4.8.

Problem 4.22: In Example 4.8, calculate the flow Reynolds number. Discuss the Reynolds number in the context of the laminar flow assumption made for this model. For Example 4.8, calculate the Reynolds number using total viscosity to be equal to the fluid viscosity and average turbulent viscosity. Define average as $(\mu_{Tmax}+\mu_{Tmin})/2$.

Problem 4.23: Repeat Example 4.9, calculating the average turbulent dynamic viscosity using the COMSOL Derived Values tool. Also calculate the total shear force exerted on the disk by the fluid.

Problem 4.24: Rerun Example 4.9 for a wider range of omega (from small to large values) and make a table of the results. Discuss the laminar vs. turbulent assumptions with reference to ratio of turbulent maximum viscosity over fluid dynamic viscosity. Also do a literature search for finding the critical Reynolds number for a swirling flow, and compare it against your results. (For reference, see *Boundary-Layer Theory*, 8th edition, H. Schlichting, K. Gersten. The transition occurs at about Re~500, according to experimental results in literature.)

Problem 4.25: Calculate the max Reynolds number for the model in Example 4.10. Also draw streamlines and study the bend effect on the flow. Calculate the "equivalent length" of the bend for having the same pressure drop.

Problem 4.26: In Example 4.10, extend the last exit part of the pipe (downstream of the bend) and find the length for which the flow becomes fully developed again.

Problem 4.27: Rebuild the model Example 4.10 for turbulence flow.

Problem 4.28: Using results given in Example 4.11, draw the contours for the y-component of the velocity vector for Re 50, 100, 400, and 1000.

Problem 4.29: Using results given in Example 4.11, draw the contours for pressure for Re 50, 100, 400, and 1000.

Problem 4.30: Run the model given in Example 4.11 and modify boundary conditions to have the two vertical side walls move in opposite directions, instead of the upper and lower walls.

Problem 4.31: In Example 4.11, calculate the maximum value of Reynolds number and investigate if the laminar flow assumption is justified. Determine the Reynolds number for which the flow becomes turbulent.

Problem 4.32: In Example 4.12, calculate the pressure change due to water hammer for a series of different boundary conditions for the pipe at the entrance.

Problem 4.33: In Example 4.12, compare the model results with those obtained by using governing equations given in the example.

Problem 4.34: In Example 4.12, make an animation of pressure variations along the pipe versus time. Use the Player button in the toolbar to create the animations.

Problem 4.35: Repeat Example 4.13 with different velocity values for Inlet. Discuss the results based on the flow Reynolds number.

Problem 4.36: Change the pressure boundary condition at the outlet in Example 4.13 and compare the results against the existing obtained results.

Problem 4.37: In Example 4.13, change the orientation of the baffles inside the mixer and run the model.

Problem 4.38: In Example 4.14, compare the modeling results with analytical $R_{th} = (r_2-r_1)/(4\pi k\, r_1 r_2)$, where R_{th} is the thermal resistance for each layer with internal radius r_1 and external radius r_2.

Problem 4.39: In Example 4.14, fill the void at the center of the sphere and assign a volumetric heat source to this region. Run the model and compare the results with previous boundary conditions.

Problem 4.40: Repeat Example 4.15 and make use of symmetry for geometry and boundary conditions.

Problem 4.41: For Example 4.16, do you recommend running this model for more than 60 seconds? Explain your answer.

Problem 4.42: For Example 4.16, calculate the dimensionless temperature quantity $(T - T_{amb})/(T_{ini} - T_{amb})$ at time equal to 5 sec. Compare the results versus e^{-BiFo}. Let $T_{amb} = (273.15 + 30)$ K and $T_{ini} = (273.15 + 450)$ K. Explain your results.

Problem 4.43: Using Example 4.16, for this model (with Aluminum wall/fin) the temperature of the fin tip reaches a steady state of about 350°C. Modify the dimensions of the fin such that the tip temperature drops to 250°C.

Problem 4.44: For Example 4.16, calculate the heat flux through the side surface of the fin and compare it to that of the tip. Discuss the results.

Problem 4.45: In Example 4.16, change the wall material to concrete and the fin material to structural steel. Estimate time constant of the problem and build the model.

Problem 4.46: Using Example 4.17, calculate the equivalent thickness for a layer of air with equivalent resistance to the thin contact resistance layer. Rerun the model using this equivalent air layer and compare the results to original solution results.

Problem 4.47: For Example 4.17, add another thin layer to the model between concrete and gypsum layers and build a new model.

Problem 4.48: Compare the results given in Example 4.18 against analytical results. The voltage for an RC circuit is given as. Show that at t = RC = 0.05 sec., the value of voltage across the resistor drops to $1/e$ % (or 36.8%) of its total value of 10 volts. Extract these results for the model and compare them. (Modeling result is 3.651 V.)

Problem 4.49: In Example 4.18, change the value of resistance to 1000, 2000, 5000, 10000 and calculate/model the circuit responses. Use Parametric sweep and resistance as a parameter.

Problem 4.50: In Example 4.19, use the analogy between mechanical mass-spring-damper and RCL circuit systems and derive the equivalent mechanical system for the model. Solve the mechanical system by hand and compare your results.

Problem 4.51: Using Example 4.19, run the model to study the effect of resistance on its behavior for R = 1000, 3000, 5000, 10000 ohm. Use the parametric sweep tool.

Problem 4.52: Using Example 4.19, run the model to study the effect of resistance on its behavior for L = 10, 100, 1000 mH. Use the parametric sweep tool.

Problem 4.53: Using Example 4.19, run the model to study the effect of resistance on its behavior for C = 1, 10, 100, 1000 nF. Use the parametric sweep tool.

Problem 4.54: Compare the stress and displacements results for the model in Example 4.20 vs. results for the model in Example 4.1. Make a table of these results to clearly show the effects of orthotropic materials on these quantities vs. isotropic ones.

Problem 4.55: For Example 4.20, use Voigt for Material data ordering and run the model. What is the difference between Voigt and Standard ordering format? (Refer to the COMSOL Manual for definitions.)

Problem 4.56: In Example 4.20, apply different loading (in plane) scenarios and discuss the results with respect to orthotropic material properties.

Problem 4.57: In Example 4.25, build a perforated plate inside the muffler and study its effects on the acoustic pressure and sound pressure levels.

Problem 4.58: In Example 4.25, build a second chamber inside the muffler to create a double-wall resonate chamber and study its effects on acoustic pressure and sound power attenuation.

REFERENCES

4.1: Maliska, Clovis R., "On the physical significance of some dimensionless numbers used in heat transfer and fluid flow," Mechanical Engineering Dept., Federal University of Santa Catarina, Brazil. (accessed, Dec. 9, 2013 at http://www.sinmec.ufsc.br/sinmec/site/iframe/pubicacoes/artigos/novos_00s/2000_maliska_encit.pdf)

4.2: Shames, Irving H. and Dym, Clive L., *Energy and Finite Element Methods in Structural Mechanics*, Hemisphere Publishing Corp, 1985.

4.3: Hartog, J. P., *Mechanical Vibrations*, Dover Publications, 1985.

4.4: Ugural, Ansel C. and Fenster, Saul K., *Advanced Mechanics of Materials and Applied Elasticity*, 5th ed., 2011.

4.5: Verteeg, H. K. and Malalasekera, W.,*An Introduction to Computational Fluid Dynamics: The Finite Volume Method*, 2nd ed., 2007.

4.6: Zhou, Y.C., Patnail, B. S. V., Wan, D. C., and Wei, G.W., "DSC solution for flow in a staggered double lid driven cavity," *International Journal for Numerical Methods in Engineering*, 57, 2003, pp. 211–234.

4.7: Ghidaoui M.S., et. al., "A Review of Water Hammer Theory and Practice," *Applied Mechanics Review*, ASME, 2005.

4.8: Cengel, Yunus and Ghajar, Afshin, *Heat and Mass Transfer: Fundamentals and Applications*, 4th ed., 2010.

4.9: Hayt, William H. et. al., *Engineering Circuit Analysis*, 8th ed., McGraw-Hill, 2011.

4.10: Keyes, David E. , et. al., "Multiphysics simulations: Challenges and opportunities," *The International Journal of High Performance Computing Applications*, 27, no. 1, February 2013.

4.11: Calkin, M. G., *Lagrangian and Hamiltonian Mechanics*, World Scientific Publishing Company, 1996.

4.12: *Encyclopedia of Polymer Science and Engineering*, Wiley, 1991.

4.13: Modeling Damping and Losses section of Structural Mechanics Module, COMSOL 4.3 manual.

4.14: *Thermoelectric Handbook: Macro to Nano*, D. M. Rowe, ed., CRC Press, 2005.

4.15 Virjoghe, Elena-Otilia et. al., "Numerical simulation of thermoelectric system," Proceeding ICS'10 Proceedings of the 14th WSEAS international conference on Systems: part of the 14th WSEAS CSCC multiconference - Volume II Pages 630–635

4.16: Jaegle, Martin, "Multiphysics Simulation of Thermoelectric Systems," Proceedings of the COMSOL Conference 2008 Hannover.

4.17: Jaegle, Martin, "Simulating Thermoelectric Effects with Finite Element Analysis using COMSOL," Fraunhofer-Institute for Physical Measurement Techniques.

4.18: Hall, Donald E., *Basic Acoustics*, California State University, Krieger Publishing Company, 1993.

SUGGESTED FURTHER READINGS

The Finite Element Analysis of Shells—Fundamentals, D. Chapelle and K. J. Bathe, Springer, 2nd ed., 2011.

Fundamentals of the Finite Element Method for Heat and Fluid Flow, Ronald W. Lewis, Perumal Nithiarasu, and Kankanhalli N. Seetharamu, Wiley, 2004.

Mechanics of Structures, Variational and Computational Methods, Walter D. Pilkey and Walter Wunderlich, CRC Press, 2nd ed., 2002.

Stability of Structures: Elastic, Inelastic, Fracture and Damage Theories, Z. P. Bazant and L. Cedolin, World Scientific Publishing Company, 2010.

The Theory of Homogeneous Turbulence, G. K. Batchelor, Cambridge University Press, 1982.

TRADEMARK REFERENCES

COMSOL® is a registered trademark of COMSOL AB.

LiveLink™ is a trademark of COMSOL AB.

SpaceClaim® is a registered trademark of SpaceClaim Corporation.

SolidWorks® is a registered trademark of Dassault Systemes SolidWorks Corporation.

Excel® is a registered trademark of Microsoft Corporation.

MATLAB® is a registered trademark of The MathWorks, Inc.

Inventor® is a registered trademark of Autodesk, Inc.

INDEX

Note: Page numbers followed by "*f*" and "*t*" indicate figures and tables respectively

A

acoustic pressure isosurfaces, muffler, 235, 235*f*
Acoustic Pressure (*acpr*) node, 234
acoustic pressure wave propagation
    automotive muffler, 234–235, 234*f*, 235*f*
        boundary conditions, 231–232
        cylinders, building, 230
        decibel type acoustic power attenuation, 237*f*
        frequency setting, 234
        geometry, building, 230, 231*f*
        mesh, building, 233, 233*f*
        model data as parameters, 229, 229*f*
        model tree for model physics and boundaries, 232*f*
acoustic sound pressure level (SPL), 228
Add Physics feature window, 26*f*
animation, modes of vibration for eigenvalues, 93, 94*f*
attenuation of acoustic power, 235, 237*f*
automotive muffler, acoustic pressure wave propagation, 228–237
axisymmetric flow, 96
    in nozzle, 96–104

B

Boolean operation, 42
Boundary condition selection, 30*f*
boundary load data entry, buckling of column, 73, 74*f*
bracket assembly
    geometry of, 54*f*, 56
    parametric analysis, 53–67
buckling of column, 67–68
    critical load, 73–75, 75*f*
        applied load, 73
        boundary conditions, 72
        control points data entry and resulting polygon geometry, 69*f*
        cross-section geometry with subtracted circular hole, 70, 71*f*
        geometry of column, 68, 70, 70*f*
        line point coordinates data entry, 69*f*
        material data entry, 72*f*
        parameters data entry window, 73*f*

C

CFD. *See* Computational Fluid Dynamics
coefficient expression
    for copper electrode blocks, 225, 226f
    for n-type TE modules, 224, 225f
    for p-type TE modules, 224, 224f
Coefficient Form for Copper Electrodes, 225
Coefficient Form for TEn, 224
Coefficient Form for TEp, 224
Coefficient Form PDE, 221, 222
    model equations, 224
Computational Fluid Dynamics (CFD)
    flow in static mixer, 150–158
contact resistance, 180–181
    parameter setup, 183f
convergence and stability, 17–20
    heat transfer in slender steel bar, 18–20
critical load of column, buckling
    analysis, 67–75
current density. *See* electric current density

D

damping, 2D bridge-support truss, 85–86, 86f, 87f
data entry interface for report document, 48f
data entry window
    for animation/player for eigenfrequencies, 52, 53f
    stress analysis of thin plate
        fixed boundary conditions for, 45f
        for Linear Elastic Material, 44f
        load boundary conditions, 46f
decibel type acoustic power attenuation, 237f
deformation of thin and thick plates, 38
discretization of fluids, 109, 110f
displacement
    of buckled column, 75f
    of thin plate, calculating, 38–48
double-driven cavity flow, moving boundary conditions, 129–141
double pendulum, multibody dynamics, 214–219

    data parameters, 216f
    geometry sketch and coordinates for, 214f
    ODEs as global equation, 216f
    variations of y-coordinates of two masses, 217, 217f
    x-displacements of, 218f
driven cavity, 129
dynamic analysis. *See also* static analysis
    and models, 95
        axisymmetric flow in nozzle, 96–104
        double-driven cavity flow,
            moving boundary conditions, 129–141
        Laminar Flow, 105–114, 119–128
        turbulent flow, 114–118
    static and, 2D bridge-support truss. *See* 2D bridge-support truss
    of thin plate, eigenvalues and modal shapes, 49–53
        density of plate, 49
        2D plot data entry for eigenfrequency, 51f, 52f
        von Mises stress, 50f
Dynamic Help, 34, 35f

E

Edge Load window, 45
eigenvalues. *See* natural frequencies
electric current density
    equation for, 220–221
    TE materials, 220
elements number for steel bar, 18, 19f
Euler-Lagrange equations, 215

F

FEM. *See* finite element method
FEMLAB, 23
finite element method (FEM)
    convergence and stability, 17–18
    formulation, 8–9
        process for, 8f
    Galerkin method, 15–16
    matrix approach, 9–15
    overview, 5–7

shape functions, 16–17
weighted residual approach, 8, 15
finite elements mesh
    double-driven cavity flow, 134, 135f, 138, 139f
    flow in static mixer, 156, 156f
    flow in U-pipe, 123, 123f
    for multilayer wall, 183f
flow velocity magnitude, 126, 127f, 128
flow velocity of pipe, 126, 127f, 129f
fluid flow, 95
fluid-structure interaction, 205
    node, 208, 209, 210
fluid velocity
    in double cavity, 135, 136f, 140f–141f
    static fluid mixer
        with flexible baffles, 210
        and pressure contours, 157, 157f
Fourier number, 173
free meshing, 31
fundamental natural frequency, 49, 51, 54

G

Galerkin method, 15–16
Geometry building window, 28f
global stiffness matrix
    procedure for assembly of, 13–14
    for triangular elements, 14–15, 15f
    2D truss analysis, 9–13
Graphics window, 25, 29
    finite elements mesh, 116f
    muffler built geometry, 231f
    stress analysis of thin plate
        for Geometry entries, 41, 41f
        with rectangle, 41, 42f
        with a rectangle and circles, 42, 43f
Ground node, 186

H

h-type convergence, 18
heat conduction, multilayer wall with contact resistance, 180–184
heat flux, TE materials, 220
heat transfer
    in slender steel bar, 18–20
    steady state
        modeling in hexagonal fin, 165–172
        modeling in multilayer sphere, 159–165
    transient. *See* transient heat transfer
Heat Transfer in Solids (*ht*) node, 169
Helmholtz type equation, 230
Help Documentation, 34
hexagonal fin, heat transfer
    convective boundary conditions, 169, 170f
    finite elements mesh, 171f
    geometry of, 166f, 168, 169f
    polygon parameters, 167f
    temperature boundary condition, 169, 171f
    temperature contours and heat flux, 170, 172f
    temperature distribution, 170, 172f

I

inlet velocity
    fluid-beam interactions for different values of fluid, 212, 212f
    of pipe, parametric values, 126, 126f
interface, COMSOL™, 24–32
internal force for member BF, 92, 93f
isotropic structural loss factor, 85, 86f

J

Joule heating effect, 220

L

Lagrangian mechanics, 214
Laminar Flow
    boundary conditions, 102, 102f
    pseudo time stepping, 134, 134f
    swirl flow around rotating disk, 105–114
    U-shape pipe, 118–129
        boundary condition, 124, 125f
Laminar Flow window
    discretization of fluid, 109, 110f
    fluid physics set up for swirling flow, 109, 109f
    k-ε turbulence model setup, 115, 116f

laminar velocity profile, 124
line point coordinates data entry, 69f
Linear Elastic Material, 44
    data entry window for, 44f
linearized buckling analysis of column, 67–75
LiveLink™ module, 27–28

M

Material Browser and Library, 30f
material data entry, 50f
    parametric analysis, 3D stress, 58, 58f
Material window, fluid properties in, 107, 108f
matrix approach. *See* variational principle approach
mesh
    static fluid mixer with flexible baffles, 210, 211f
    steady state heat transfer, 163, 164f
    stress analysis of thin plate, 46
mesh data entry window, 47f
mesh parameters, transient flow, 147, 147f
mesh resolution, double-driven cavity flow, 136
meshing, 29, 31, 31f
    COMSOL feature, 23
modal shapes of thin plate
    animation of, 52
    calculating, 49–53
Model 1 (*mod1*) node, 215
Model Builder window, 25, 29, 31
    advanced physics options, 115f
    work plane setup, 119, 119f
model building, 24, 25
    guidelines, 34–36
    ways for, 27
Model Library, 33–34, 34f
    3D stress analysis, 53
Model Wizard window
    for adding Laminar Flow physics, 97f
    stress analysis of thin plate
        for selecting and adding physics, 39–40, 40f
        for selecting space dimension, 39, 40f

modeling
    complex and multiphysics problems, 193–194
        acoustic pressure wave propagation, automotive muffler, 228–237
        double pendulum, multibody dynamics, 214–219
        orthotropic thin plate, 194–197
        static fluid mixer with flexible baffles, 205–213
        thermal stress analysis of bucket, 197–200
        for thermoelectric modules, 219–227
        transient response of bucket, 200–205
    COMSOL features, 23–24
    heat conduction, multilayer wall with contact resistance, 180–184
    RC electrical circuits, 185–188
    RLC electrical circuits, 188–193
    steady state heat transfer
        in hexagonal fin, 165–172
        in multilayer sphere, 159–165
    transient heat transfer, nonprismatic fin with convective cooling, 173–180
modules, COMSOL™, 32–33
muffler, automotive, acoustic pressure wave, 228–237
multibody dynamics, double pendulum, 214–219
multilayer wall
    finite elements mesh for, 183f
    geometry, 182f
    temperature and heat flux, 184f
    temperature variations, 184, 184f

N

n-type TE modules, coefficient expression, 224, 225f
natural frequencies, calculating
    thin plate, 49–53
    3D truss tower, 92–93
    2D bridge support truss, 81–83
natural fundamental frequency, RCL circuit, 189

Navier-Stokes equations, 8
  Laminar Flow, 106
nodes number, for steel bar, 18, 19f
nozzle
  geometry and dimensions of, 96f
  geometry and line segments coordinates for, 99f
  geometry cross-sections, 99f, 100, 100f
  inlet velocity
    parametric sweep values, 102–103, 104f
    surface velocity, 103, 104f
  material properties for, 100, 101f

O

ODEs
  heat transfer in slender steel bar, 18
  nonlinear system of, 215
  modeling RLC electrical circuit, 188–189
  solving, COMSOL feature, 24
orthotropic Kevlar-epoxy, material properties for, 194, 195t
orthotropic thin plate, stress analysis, 194–197

P

p-type convergence, 18
p-type TE modules, coefficient expression, 224, 224f
parameters data entry
  buckling of column, 73f
  parametric analysis, 3D stress, 55f
parametric analysis
  static fluid mixer with flexible baffles, 210
  3D stress, bracket assembly, 53–67
    applied load, 59, 60, 62f, 64, 65, 66f, 74
    boundary conditions, 59, 59f, 60f
    coordinates, data entry for, 57f
    deformed bracket geometry, 62f
    importing CAD file, 56f
    Principal Stress Volume, 62, 63f
    Study Extension data entry window, 65f
    variables, data entry for, 56, 57f
    volume Integration data entry window, 64f
    von Mises stress, 60, 61f, 62, 62f, 66f

Parametric Sweep data entry, 74, 75f
partial differential equations (PDE)
  Coefficient Form of, 221, 222
  solving, COMSOL feature, 24
  for transient heat transfer, 173
  and weighted residual approach, 15
PDE. *See* partial differential equations
Peltier effect, 220, 220f
pipeline geometry, transient flow analysis, 144–145, 145f
Poisson's ratio, 44, 195
Pressure Acoustic Model 1 node, 230
Pressure Acoustics, Frequency Domain (*acpr*) node, 231
pressure wave propagation, speed of, 142, 144
Principal Stress Volume, 62, 63f

R

RANS. *See* Reynolds Averaged Navier Stokes equations
Rayleigh Damping, 85
Rayleigh damping method, 201, 201f
RC circuit nodes, data, 186t
RC electrical circuits, modeling, 185–188, 185f
reaction forces, parametric 3D stress analysis, 65, 66f, 67, 67f
Reynolds Averaged Navier Stokes equations (RANS), 114
Reynolds number, 95
  calculating, static mixer, 157–158, 158f
  double-driven cavity flow, 130, 140, 141f
  turbulent dynamic viscosity and, 118, 118t
  turbulent flow, 114
Reynolds stresses, 114, 117
Rigid Connector boundary conditions, 202, 202f
RLC electrical circuits, modeling, 188–193
  parametric analysis for effect of capacitance on voltage output, 191–193
  and reference nodes, 189f
rotating disk

Laminar Flow
    angular velocity, 106
    axisymmetric geometry and
        dimensions for flow, 107, 107f, 108f
    boundary conditions, 111f
    swirl flow around, 105–114
    3D and 2D axisymmetric geometries
        for flow, 105, 105f
    turbulent flow, swirl flow around, 114–118

S

Seebeck effect, 219, 220f
Select Study Type window, 26, 27f
shape functions, 16–17, 19
slender steel bar, heat transfer, 18–20
SolutionMeshFiner, 138, 139f
sound pressure, automotive muffler, 234, 235f
SPL. See acoustic sound pressure level (SPL)
stability, convergence and, 17–18
    heat transfer in slender steel bar, 18–20
static analysis
    and dynamic analysis, 2D bridge-support
        truss. See 2D bridge-support truss
    thin plate, stress analysis, 38–48
static fluid mixer, 150
    boundary conditions, 154–158
    with flexible baffles, 205–206
        boundary conditions for flow, 209, 209f
        discretization of fluids, 210
        displacements of baffles, 212, 213f
        fluid velocity, 210, 212, 212f
        geometry of mixer, 206–208, 206f, 207f
        material properties, 208, 208f
        parametric analysis, 210
        von Mises stress, 210
    fluid properties, 153–154
    geometry of, 151–153, 151f–153f
stationary window, parameter values for disk, 112, 112f
steady state heat transfer
    modeling in hexagonal fin, 165–172
    modeling in multilayer sphere, 159
        boundary conditions, 162–163

materials and properties, 162, 163f
model geometry building, 161
parametric curve data for half-circle, 160, 160f
temperature distribution, 159, 164f–165f, 165
2D axisymmetric geometry, 161f
stiffness matrix, global. See global stiffness matrix
stress analysis of thin plate, stationary loads, 38
    Boolean operation, 42
    data entry interface
        for report document, 48f
    data entry window
        fixed boundary conditions for, 45f
        for Linear Elastic Material, 44f
        load boundary conditions, 46f
    Graphics window
        for Geometry entries, 41, 41f
        with rectangle, 41, 42f
        with a rectangle and circles, 42, 43f
    mesh data entry window
        and resulting mesh, 47f
    Model Wizard window
        for selecting and adding physics, 39–40, 40f
        for selecting space dimension, 39, 40f
    von Mises stresses, 46, 48f
stress analysis, orthotropic thin plate, 194–197
stress-strain relationship, orthotropic material properties, 195
stress tensor, 54
Structural_Mechanics_ModuleTutorial_Models, 56
Study and Results window, 32f
Study Extension data entry window, 65f
surface velocity
    in nozzle, 103, 104f
    3D, for values of omega, 112, 113f, 114f
    turbulent flow, 117, 117f
Swept tool to build volume elements, 233
swirl flow. See also specific flows

around rotating disk
    Laminar Flow, 105–114
    turbulent flow, 114–118

T

TE modules. *See* thermoelectric modules
Thermal Linear Elastic Materials 1 node, 199
thermal stress analysis of bucket, 197–200
    boundary conditions for, 199, 200*f*
    von Mises stresses, 199, 200*f*
thermoelectric (TE) modules
    coefficients for n-and p-type, 224, 224*f*, 225*f*
    efficiency of, 220
    heat flux and electric current density, 220
    multiphysics model for, 219–221
        boundary conditions during model construction, 221–227
        coefficient for copper electrode blocks, 225, 226*f*
        geometry of materials, building, 222, 222*t*, 223*f*
        parameters for material properties, 222, 223*f*
        thermoelectric power generation, 225
        voltage variations, 227, 227*f*
thin plate
    dynamic analysis. *See* dynamic analysis
    stress analysis. *See* stress analysis
Thin Thermally Resistive Layer, 181
3D stress analysis, parametric, bracket assembly, 53–67
3D truss tower, static and dynamic analysis
    boundary conditions, 90
    forces and stresses, 91, 91*f*
    geometry of truss, 89, 89*f*
    joint coordinates for, 88*t*
    material properties, 90, 90*f*
    modes of vibration for different eigenvalues, 93
    natural frequencies, calculating, 92–93
    normal forces of truss, 91–92
    variables for normal forces, 90

transient flow analysis
    boundary conditions and pipe shape and dimensions, 145
    water hammer model, 142–150
transient heat transfer, nonprismatic fin with convective cooling, 173–174
    boundary conditions, 176, 177*f*
    parameters setup and final geometry, 175*f*
    temperature distribution, 178, 178*f*
    temperature variation, 179, 179*f*, 180*f*
transient response of bucket
    applied load, 202
        displacement components, 204, 205*f*
        von Mises stress, 204, 204*f*
    dynamic modeling, 200–205
    Rayleigh damping method, 201, 201*f*
    Rigid Connector boundary conditions, 202, 202*f*
    solver parameters setting, 203*f*
transient solver parameters, 148, 148*f*
triangular elements, global stiffness matrix for, 14–15, 15*f*
turbulent dynamic viscosity, 117, 117*f*
    and Reynolds number, 118, 118*t*
turbulent flow, swirl flow around rotating disk, 114–118
2D bridge-support truss, static and dynamic analysis
    boundary conditions, 79, 80*f*
    geometry of truss, 77–78
    harmonics, displacement for range of, 83–85
    material properties of truss, 79*f*
    meshing, 80–81
    natural frequencies calculation, 81–83, 83*f*
2D truss, 9*f*
    analysis of, 9–13

U

U-shape pipe, Laminar Flow, 118–129
Unit System options, 29*f*
unsteady heat transfer. *See* transient heat transfer

V

variational principle approach, 8
    global matrix assembly, general procedure for, 13–14
    global matrix for triangular elements, 14–15
    2D truss, analysis of, 9–13
version 4 of COMSOL, 23–24
Vin (Fluid velocity), 210
voltage source data and nodes, 191f
volume Integration data entry window, 64f
von Mises stress, 54, 66f, 196
    dynamic analysis of thin plate, eigenvalues and modal shapes, 49
    parametric analysis, 3D stress, 60, 61f, 62, 62f, 66f
    static fluid mixer with flexible baffles, 210
    stress analysis of thin plate, stationary loads, 46, 48f
    thermal stress analysis of bucket, 199, 200f
    transient response of bucket, 204, 204f

W

Wall window, boundary conditions set up, 110, 111f
water hammer model, transient flow analysis, 142–150
water hammer phenomenon
    pressure pulse for, 142, 148–149, 149f
    pressure variation, 149, 150f
weak formulation for FEM, 19
weighted residual approach, 8, 15

Y

Young's modulus, 44